Hume

Rousseau

Kant

Hegel

Descartes

Marx

McKaneism
The Foundation of Humanity

More

Engels

Washington; Jefferson; Roosevelt; Lincoln

Aquinas

James

Lenin

Aristotle

Plato

Socrates

by Frank P. McKane

?

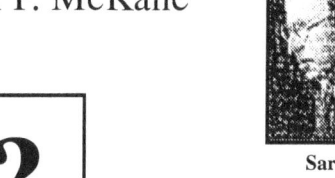
Russell

Sartre

ISBN: 1-4140-0024-3 (e-book)
ISBN: 1-4140-0025-1 (Paperback)

Library of Congress Control Number: 2003095503

This book is printed on acid free paper.

Printed in the United States of America
Bloomington, IN

1stBooks – rev. 08/29/03

McKaneism
The Foundation of Humanity

by Frank P. McKane

First Edition (Unedited, Uncensored)

Frank P. McKane

I dedicate this book to Humanity, "WHO" never stopped allowing me to observe, wonder, dream, think and conclude on all the fantastic things and events that surround us. Very special thanks for my son Frank, who had to experience first hand, the closeness and full impact of my thinking mind with all the ups and downs throughout my growing and personal development process. Also, lots of thanks to the contribution of all the people around me who I ran into in my life and all those whose thought processes, behaviors, and actions influenced, formed, and enriched my knowledge and enabled me to construct, think through, and summarize the philosophy that you will discover in this publication.

Table of Contents

Table of Contents

Preface

"The Foundation of Humanity" - a tall order to claim, that I am the one in the history of the 100,000 years of existence of "Homo sapiens" who can define the foundation for our human existence. However, that is what this book is all about. To understand the place of this work in human history, I have to clarify a few things here at the beginning.

For easier understanding I can compare my work to building a house. I have no intention to tell the whole human race how to live, or which set of rules they form their societies by, but in the same extent, like building a house, the foundation has some defining rules as to how the rest of the house can be built. For anybody who thinks I set or designed this "Foundation of Humanity" the answer is simple: the foundation coming from a more or less objective observations of our existence. So build your three or five bedroom house or even your castle, the only thing I want to ensure is that your foundation doesn't get built on "quick sand" or it doesn't have any "serious cracks" in it, all which could bring down the building on your head and destroy you.

My observations will show you later in this book how many cracks we have in our current foundation as I am writing this book in the year 2003.

This work is Philosophy that I try to gain the respect and acceptance for this science (my passionate pastime in the last thirty years) that many times was ridiculed with all the authors included, who tried to make a difference in our life, for the better.

This work attempts to be relatively simple in Part 1, so all readers with basic schooling and a little extra reading can easily understand the thoughts and reasoning in Part 1 of this book.

I will refer to names and their thought processes in the book from time to time and this may require the readers to extend their reading with a few other publications for better understanding. In Part 2, I intend to go deeper in the world of philosophy and it will require a lot more extended reading and study from the reader to understand my arguments.

The word "Unedited" means, that I try to publish my original thoughts only edited for English lightly. I will address these facts in a later chapter when we will analyze the human mind and the way it works. "Uncensored" means that I didn't let anybody for any reason alter the contents of this book.

Preface (cont.)

I hope you will enjoy your reading experience, and by the end of this book, will have a little more open mind to view the world around you.

Finally, I can list pages of names here, to thank all of those who helped my journey to get to this point in my life, in my thought process and also in the possibility to publish this work, but instead of that I dedicate this book for all of them in the special page that you already read at the beginning. One exception, a special thanks for Kent Meiser who helped me with publishing and for believing me as my friend and my long time business partner.

I wish you a very enjoyable journey reading this book, and I hope to accomplish the introduction of all those thoughts for you, that I find as a very important reason to write this book.

Frank P. McKane

Part 1

Frank P. McKane

Part 1 - Introduction

Philosophy vs. "Set of Views and Experiences", Philosophy is the Science of the Sciences

Philosophy vs. "Set of Views and Experiences"

McKaneism - The Foundation of Humanity. It sounds very serious. It is!

This is an undertaking to summarize the state of mind of humanity in the early 21st century, put philosophy where it is belongs, and try to set directions for the future of humanity. This is philosophy that is intended to discover, analyze, and introduce "The Foundation of Humanity".

In this point – "living in America" (USA) – this simple question applies:

"Who the hell do you think you are?"

Before I answer this question I have to take a very short tour in the history of philosophy. This tour will answer some questions about the current "state of philosophy".

The last time in human history when real philosophers existed, it was about 2500 years ago in the Greek city of Athens, then later in Rome. The reason for this, is because during that time the social structure was "Slavery" so the conquering powers used the conquered people as slaves to build their roads, grow their foods, and serve them in their households. (I put "Slavery" between quotes because it represents a social structure in human history, more about it later.) So, the so called citizens of those nations had a free life on account of the slaves. This way they could choose what they wanted to do with their life, because the goods and services of basic living was provided for them. They could pick up the subject of philosophy and meditate and think about the thought processes, but still not too many people paid attention to them.

There were existing, walking, thinking entities in the society, and only a few of them remembered by human history.

Even during that time the danger of philosophical thinking arrived, when one of the famous philosophers, Socrates, was critical about the government, and he ended up dead. But some of the others managed to put down the foundation for philosophy, like Plato and Aristotle. (I can't bypass a very serious observation here, that "Slavery" is a non-working social structure that was proven poor 2500 hundred years ago, and how smart humanity is, that we try slavery in many other countries later in human history.

5

Part 1 - Introduction
Philosophy vs. "Set of Views and Experiences", Philosophy is the Science of the Sciences

Philosophy vs. "Set of Views and Experiences" (cont.)

I guess we (humanity) learn very slowly.) Moving on in history, the next social structure, "Feudalism", followed where the main philosophy became a fate based (mostly Christian) philosophy. It showed us a closed-minded humanity that almost executed Galileo, who argued that the Sun and not the Earth was the center of "our universe". He had to revoke his words to stay alive — alone, ridiculed, and watched for rest of his life for "good behavior". The next big time discredited philosopher was Marx in "Capitalism", and then Lenin statue fell to the ground with the falling of "Communism". This is a crash course of history through 2500 years with mentioning only a few of the most famous names, but there were many others. The interesting observation is that all of these people were "part time" philosophers after the Greek time.

Most people in philosophy were scientists that picked up philosophy as a "part time" interest. Here is my observation and explanation for that: I am very sure that up to this book, no one truly placed philosophy where it belongs.

Now, I will answer the question:

"Who the hell do you think you are?"

The answer is very simple:

"I am Nobody", another part time philosopher, with a very strong passion that for the first time in human history, I will place philosophy where it belongs, or at least I will start the process! BUT, "I am a very special Nobody", I am someone who had the Worst and Greatest fortune at the same time. I lived half of my life in the "Communist" society and the second half in the "Imperialist" society. Communism is considered the biggest evil on Earth by most people without really knowing anything in detail about it. Imperialism, by some definition is the final, most advanced phase of Monopole Capitalism, and we can say - "as good as it gets"! If I managed to puzzle you with my definitions of these terms of philosophy, please don't be discouraged. We will get into these subjects and definitions in great details later in this book.

Being "Nobody" in this society, gives me a great advantage to express my views without any major consideration as to who will like it or who will not like it.

Part 1 - Introduction
Philosophy vs. "Set of Views and Experiences", Philosophy is the Science of the Sciences

Philosophy vs. "Set of Views and Experiences" (cont.)

I am not afraid to bring up any subject, or analyze any thoughts or issues or processes, or even being right or wrong. Next to that I have a gift that only a few people had that I encountered in my life — I am a great observer. I can observe things in life that can be greatly overlooked by most people. Using these capabilities, I just know somehow that I have to write down these thoughts and observations. This publication is priceless for Humanity, because it will open the door to the next millennium or more for all the following free thinkers who will take this publication as an example. They may build on it or may depart from it and I am sure they will come up with much more than I ever dreamed of.

You can take my word for it: This is just a beginning of the realization of the importance of Philosophy as a science in our existence. If you look at Philosophy very closely, you will realize that more people's lives were lost or changed for standing up or believing in a "Specific Philosophy" than any A-bomb or other technical invention. The collision of Humanity always can be tracked back to the collision of "Thoughts or Philosophies" before the confrontation begins with arms throughout human history.

Before I forget, "McKaneism", the word I came up with, is nothing else except just to say that this is a "Philosophical Thinking of Frank P. McKane". I like the fact that I have these thoughts and this is why I named it after myself like a father names his son, but most important, it is the organized format of this information's thoughts, conclusions, and the fact that it is a "System of Thoughts" or "Philosophy". That is most important, so from here, one can name it whatever one wants to call it.

Philosophy! We all have our own Philosophy! True? False?

The answer is: It is not true!, It is false! We all think we do, but we don't! It already sounds like we are playing with words but that is not what we are doing. Throughout this publication you will encounter numerous times when it will look like we are running around with words, but we don't. We just have to be real precise to address every issue and to define our words, what they mean and what we try to communicate using them. It is very important in philosophy to find starting points, build logical thought chains to follow from one step to the next, until we reach our conclusion.

Part 1 - Introduction
Philosophy vs. "Set of Views and Experiences", Philosophy is the Science of the Sciences

Philosophy vs. "Set of Views and Experiences" (cont.)

This way every reader can have a chance to journey with us and approve or disapprove the process of our chain of thinking and the final conclusion.

So, we all don't have our own Philosophy! However, we all have our "Set of Views and Experiences". These are the thoughts of information, knowledge, and experience that we learn and acquire going through life. Philosophy is a scientific system, a complete and orderly, organized, and comprehensive "Set of Views". For an easy to understand example: if one takes the American English alphabet, then A to Z is "Philosophy", while saying A,B,C,X,Y,Z is only a "Set of Views". Using another example: When we go out shopping for clothing or buying a gym shoe, we are searching for something we like, something that grabs our attention, fits comfortably, yet I don't know anybody who will stand in the middle of the Mall and start to analyze the "Aesthetical Principles of Philosophy and Action" of buying a shoe. This second example also shows that one can live a comfortable life and can make most of the proper decisions (but not all!) with the proper "Set of Views and Experiences" without extending them to a complete Philosophical System.

For another example: Christianity, Islam, Judaism, Buddhism, Taoism, Marxism, Existentialism (just to have a few examples) are Philosophies. However, the smart advice of the corner store owner or some of our friends' conclusions of their life knowledge and experience, or in other words their "Set of Views and Experiences", (this does not mean that one can't learn from it), simply that it isn't a Philosophy.

I will use throughout this publication a lot of examples, because some of them can create a picture in our mind, that can describe our intended point a lot more precise way than writing 5 to 6 pages of information about the same subject. (As someone once said "A picture is worth a thousand words".) Most of the analysis, examples, and explanations will come from the USA. I have to repeat many, many times that the USA, the technically most advanced society of the 21st century — I came to the conclusion that the USA is in the final most advanced stage of Capitalism that exists on Earth.

That leads us to the next very important question: If this is the final stage, how long can it go on? What is the next possible economic social structure after Capitalism?

Part 1 - Introduction

Philosophy vs. "Set of Views and Experiences", Philosophy is the Science of the Sciences

Philosophy vs. "Set of Views and Experiences" (cont.)

Is it better that we start working on the creation and understanding of the next social format, or just let it come alive as all the others did before?

Here when I say "social", I am not talking about "Socialism". I am simply using the word in the loose interpretation of our daily language, that when people get together to perform activities like after hours interaction with each other, we call it "socializing". When people of their own country create or declare their own Constitutions, they form a "social format" they want to exist within, but that does not have to be "Socialism"! I had to clarify this here so any of my readers don't get the impression that this is a Socialist philosophy manual. It is not!

We didn't even begin any serious discussion and you can see how many times I had to detour my introduction to try to properly describe what I am trying to say.

That is one reason why philosophy of any kind is a difficult reading because the author spends most of his or her time (of course if the publication is real serious about philosophy!) defining words and meanings of those words that they are using. It is very important, because this is the only way to really communicate between human beings, that if the same word has the same meaning for everybody who is in the discussion. This is why philosophy books end up being six, eight hundred pages or more, analyzing only a few topics.

Many times I use "for example" the USA for better understanding of some of my statements, without giving any lengthy explanations about it. In the later chapters we will spend a great deal of time analyzing "possible social structures", including the USA. At that time I will try to explain, compare, and introduce most of the existing "social structures" on our planet in the beginning of the 21st century and the beginning of time as we understand it based on our knowledge of today's science.

In the same manner I made some very strong statements of the subject matter "Philosophy vs. Set of Views and Experiences" without getting into any serious discussion about the issues. We will do that in great detail later.

Part 1 - Introduction
Philosophy vs. "Set of Views and Experiences", Philosophy is the Science of the Sciences

Philosophy vs. "Set of Views and Experiences" (cont.)

Before we can move on to Chapter 1, two more things I need to introduce. Just like from the beginning of this book, I will make some statements and presentations here to describe the structure and the flow of thoughts of this book. In the later chapters I will revisit all these statements and charts and I will give some more serious explanations and backing up my words with some serious proof of observations or conclusive thought processes.

The first thing I introduce is the relation of philosophy to other sciences, then the second is the structure of the new system called McKaneism — that describes the complexity of matter and mind, how they relate to each other, how the complexity of the two are structured together as a system, and how it is my observation that helped me discover this system, that I believe exists in the real world with or without human existence being involved in it.

Philosophy is the Science of the Sciences

This is a very serious and overlooked fact in our current societies in the 21st century. Most people don't pay attention or ignore the importance of Philosophy! Some times in some societies on Earth the leading people of that society help to ignore philosophy for the others so a lot less people thinking, a lot less problems they have. (This is one of many things they are wrong about!)

We try to follow a flow of logic in this book to build a simple to complex road for all our travelers. That will allow all of us to travel together in our discovery lane and think together on all the subjects. I will try to use realistic thinking as a method to reflect on all the issues that will be introduced in this book. If you will feel from time to time that the topics are difficult, don't be discouraged or surprised, because they are really difficult!

Chart 1 is the introductory picture to try to visualize the connection between Philosophy and all the other Sciences: (of course not all listed!)

Part 1 - Introduction
Philosophy vs. "Set of Views and Experiences", Philosophy is the Science of the Sciences

Philosophy is the Science of the Sciences (cont.)

Chart 1

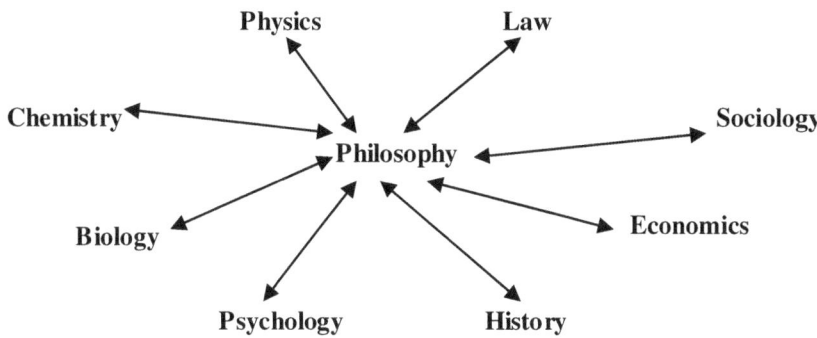

Philosophy is the Science of the Sciences. It does not mean that philosophy is superior, but it shows us that philosophy connects all the sciencies together. It also shows us that each science is influenced by philosophy at every point of time and philosophy responds with challenges and approvals to direct, guide, and accelerate the sciences. Philosophy as a science should and does connect and guide the other sciences.

This interaction exists from the beginning of time of "Human History", regardless if we understand it, accept it, or deny it by any human society. This is one of the very important discoveries of McKaneism.

This observation leads us to the very important conclusion that only philosophy can explain the way all human knowledge is connected, and lead us to the universal explanation of the questions of humanity and intelligent existence.

Let us think through a few small examples to put our previous Chart 1 in a more focused perspective. For example, in our current 21st century we arrived at the problem of human cloning.

Our science of biology and medical science was able to understand most of the genetic sequence of our human DNA. So here we are ready to play with this new knowledge like kids play with their new toys.

Part 1 - Introduction
Philosophy vs. "Set of Views and Experiences", Philosophy is the Science of the Sciences

Philosophy is the Science of the Sciences (cont.)

How does Philosophy come to this picture?

In our current time about 95% of the Earth population has a philosophy that is based on one or another form of religion, and the belief of a higher power that governs our life through the teaching of "The Book" and miracle signs. ("The Book" can be like the Bible or the Koran, etc.) In these Philosophies we should ask the higher power: How can we deal with "Human Cloning"? But it is not in "The Books", because at the time the books were written, "Human Cloning" wasn't an option for humanity, and so the knowledge was and is non-existing!

Another problem is that the higher power doesn't answer at all! So, people who represent the higher power try to give us the answers. For me it is unacceptable. In my view they are just as clueless as I am about the answer, and they definitely didn't hear an answer from their form of a higher power! So they make up their version of moral code, and try to apply it for all of us, because they simply don't have any answer at all! So as we see Philosophy affects Biology and Medical Science and Biology and Medical science tests the validity of those Philosophies. Just like you find the arrows pointing in both directions in our Chart 1, between the sciences and philosophy. Just to peak your interest, McKaneism has a valid, correct answer for this issue and you can read about it later when we address the functioning ways and principles of our human mind — that drives any conscience, thinking mind.

Another interesting example is the issue of abortion. Another area where philosophy and medical science and sociology keep influencing each other to try to find the proper answer or answers for this issue. Again you can find the connecting arrows in my Chart 1.

The final set of thoughts in this introduction are the representation of McKaneism with Chart 2, which describes the way this philosophy organizes matter and mind and all their existence, as well as rules around them into one system. I will introduce you to this chart in the next page and in the following chapters I will spend a great deal of time to open up and explain the thought process around each and every level of this representation.

Part 1 - Introduction
Philosophy vs. "Set of Views and Experiences", Philosophy is the Science of the Sciences

Philosophy is the Science of the Sciences (cont.)

Let us take a quick look at Chart 2 and spend a short time now to review the elements of this chart, then I will move on and start the first chapter going into the analysis of the philosophy I called McKaneism. Chart 2 is one of the Basic organizational charts of McKaneism. Let us move on for now and start investigating my philosophy in greater detail

Chart 2

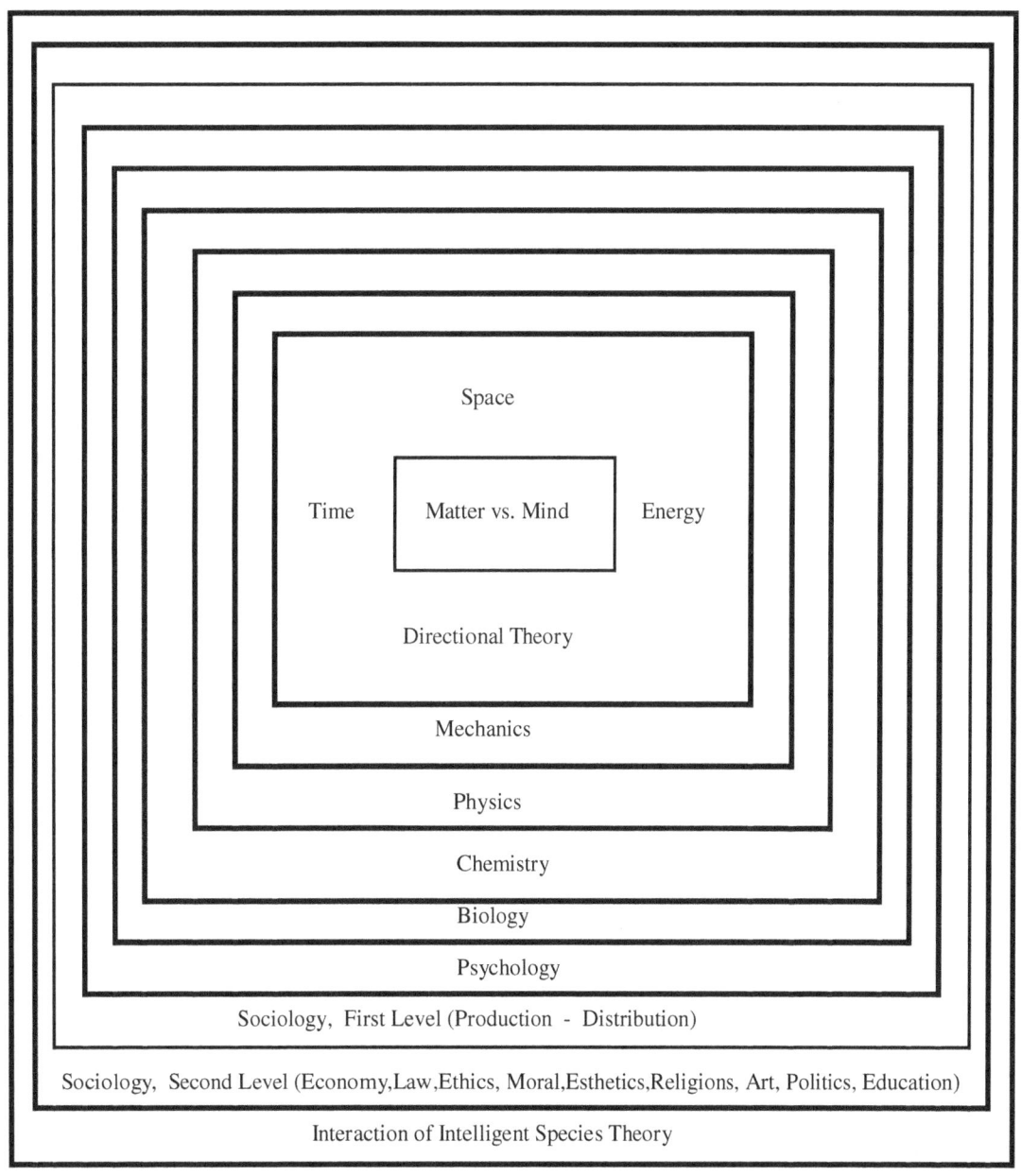

Space

Time   Matter vs. Mind   Energy

Directional Theory

Mechanics

Physics

Chemistry

Biology

Psychology

Sociology, First Level (Production - Distribution)

Sociology, Second Level (Economy,Law,Ethics, Moral,Esthetics,Religions, Art, Politics, Education)

Interaction of Intelligent Species Theory

Part 1 - Chapter 1
Basics

Basics

Here we are at the beginning of the 21st century and Humanity is still in the early time of their existence as far as philosophy is concerned. To make it easy to understand: if you compare the life of philosophy as a science to a life of a human being, philosophy is in the teenage years of the human being existence — about 16, 17 years old, by maturity.

As you can see, I made a statement that anyone can challenge by saying "How do you want to prove your statement?" As you realize, the difficulty of proving any statement, you must also realize that when we are speaking we deliver our thoughts to the listener. Our thoughts have a pattern to form in our mind, based on information we accumulated and stored throughout our life.

The base for most philosophy is to form thoughts, and explain those through the definition of the meaning of the words and phrases, and use them to build a "System of Thoughts", that describes the specific subject matters.

You will discover it later as we go that it is not an easy task to do. It is very important to be precise with our definitions so we all can understand the subject matters, review the statements, debate the validity of each and every one of them. Another example to make understanding easier. If I say 2+2=4 we all know what am I talking about. If you take any object on Earth, for example the pen, and take 2 pens + 2 pens and you put it on the table you will end up with 4 pens on the table. To prove it you can look at the table and count 1,2,3,4 and count all 4 pens on the table. Case closed, 2+2=4 is proven based on your own observation. For those who learned mathematics "seriously" (a lot more than adding and subtracting) it is a mathematical proof through reasoning in the mind that takes you to the same conclusion.

Those of you who try that may have discovered that it isn't as simple as counting pens on the table. In this case we had both the observation and the reasoning thinking available to get to the same conclusion.

We are not that lucky with all the subject matters around us.

Part 1 - Chapter 1
Basics

Basics (cont.)

For example when we get down to the atomic level of the matter we can't observe them with our senses (eyes), so we build instruments to help extend our capabilities to observe them, and also we use very heavily our reasoning thinking through the language of mathematics.

Even then we have limitations how far we can go. For example we have a "Big Bang" theory that we believe happened 15 billion or so years ago. We have no way to observe it as we know today in the early 21st century, so we use very complicated mathematical formulas to prove our theory. (When I say "our" in here I refer to Humanity. Humanity has a knowledge base for all who willing to study long hours to acquire the necessary mathematical knowledge. I don't try to include or exclude anyone at this point, including myself, to believe or not in this theory.) Through this example we can see that the more complex the issue that we are analyzing, the more difficult to debate, explain, describe, and conclude on any subject matters.

Now you know that I will use words and phrases to describe the thoughts and theories but what will be a subject matter? The subject matter will be the original, historically viewed subjects of philosophy and I will extend on to create a new philosophy that I called McKaneism. The name simply represents that this philosophy is "The System of Thoughts of Frank P. McKane". The brave part is coming when I extend my vision to call my philosophy "The Foundation of Humanity".

Only time and history will tell that my observations and vision is really what I think it is. As I stated before, the philosophy always went hand to hand with the other sciences just like I represented in Chart 1 of the introduction pages.

However, as of today we have many schools of thought and philosophy that actually contradicts or disregards each other. For example if one believes in "Christianity's Creation Theory" then that one probably denies the "Darwin's Evolution Theory" because the two have contradicting observations, reasoning, and end with the opposite conclusions.

Part 1 - Chapter 1
Basics

Basics (cont.)

I will present a comprehensive approach to organize and place all the existing human knowledge in one system, even trying to analyze and resolve the conflicts of different philosophies, placing them in the proper place and time in the human sciences and history.

Philosophies throughout the history of humanity try to answer a few very basic questions such as :

Who are we (humanity)? What is our (humanity) origin? What is our surroundings (universe, space, time) that we observe? What is our conscience mind? Why are we the only ones (so far as we have observed) with a conscience mind and the capability of thinking? Why are we so different from other existences (animal world) as we know them? What are the principles and answers for our social behaviors and social structures?

These are the most basic questions. The answers for all or some of those questions define what kind of philosophy we are analyzing. If you think the questions are easy you are mistaken. Humanity try to answer some of those questions in the last ten thousand years of our history.

The in depth investigation may or will surprise you on how much some of the theories overlap each other, but none of them gives us a correct or reasonably acceptable answers up to today. All the philosophies build a thought process and try to start from a few accepted points or principles and try to build a reasoning thought stream to prove their validity in life. As we will see in closer investigation, they all overlapping some points, competing with each other, trying to discredit or disprove each other.

It is time to set the record straight about the existing "Systems of Thoughts". As we said that before many schools of thought exist in the World currently in the beginning of the 21st century. All these thought processes try to answer all those questions that we read about on this page. If we uphold our statement that "Philosophy is the Science of the Sciences" **all** the previously developed "Systems of Thoughts" or call them "Philosophies" **will fail us**.

So, where are we now? Until now, in this publication I was generous to all those so called "Philosophies" that were developed throughout the history of humanity.

Part 1 - Chapter 1
Basics

Basics (cont.)

Now I will put them in the right place and time without providing too much proof of my statements here. You will feel that it makes sense what I am saying and that is all I want to communicate now. In the later chapters, I will revisit all these thoughts and I will prove it, why my statements are correct here.

With the deepest respect from the bottom of my heart, I have to admit that I always have been fascinated with the thoughts and thinking of all those people who dedicated their life to enter this very unrewarding world of science called "Philosophy".

Most of them were scientists in the fields of mathematics, physics, architecture, engineering, literature, sociology, etc. Just a few names like Aristotle, Plato, Galileo, Hegel, Marx, Sartre, and of course many, many others. These great individuals put down the foundation of thinking for humanity. However, in their time most of them entered the field of philosophy to extend their search of knowledge. For example: Marx was analyzing the social structures of the existing society in his time when he developed his theory of communism. He was mostly a political activist with strong interest in social science that we call sociology today. For example, history records that one time someone asked him about his thoughts, referring to it as Marxism. He replied: "If that is the way you want to call it". He knew he had a "System of Thoughts" but even he himself didn't name it. Later in history, because he attracted many followers like many other schools of thought, history named his "System of Thought" Marxism.

So, to summarize it for now, in their time in history their thoughts were "philosophies", as of today in the time of our history at the beginning of 21st century I will call it "Systems of Thoughts". Philosophy as a science of sciences should organize all the existing sciences in one round picture, and grow and change with them as humanity is changing through history. None of these previous "Systems of Thoughts" were able to accomplish that.

Here is a bold statement : McKaneism did it!

My greatest wish to accomplish in this book, and (by reading and thinking with me about all the subject matters,) I hope all of you will find a proof at the end, that:

17

Part 1 - Chapter 1
Basics

Basics (cont.)

"McKaneism as a Philosophy ("The Systems of Thoughts of Frank P. McKane") builds on observations and the knowledge and discoveries of sciences and all the predecessors and their "Systems of Thoughts", and will be "The Foundation of Humanity" for a long time to come.

Let us take a look — first the outline of McKaneism, the area of things and thoughts that I will cover, then I will build a "New System of Thoughts" and we can see if it will be or won't be able to stand up against all the questions and scrutiny that McKameism will and should encounter from humanity.

Finally in Chapter 1 let us go through an outline of the new philosophy McKaneism. Let us see the components of this philosophy.

Beginning with analyzing the: "Fundamentals as Matter vs. Mind, Space, Time, Energy". Up to this point the subject matters are not new and many other "schools of thoughts" did it. What is new is the way we will define each of them slightly departing from the past. However the **"Directional Theory"** is a new addition to the fundamentals and the way I will organize them, also will be new.

Moving on to build our thought system we will analyze: "Mechanics, Physics, Chemistry", "Biology", and another new addition the **"Attributes of Living Entities Theory".** "Physiology", going around the human mind, and something new again the **"Knowledge Theory and Chart"** and the **"Attributes of Human Beings Theory"** which is the extension of the "Attributes of Living Entities Theory".

Then I will enter the world of societies and I will analyze: "Sociology", Structures and value systems of societies, many of the previous "Schools of thoughts". Another new thing I introduce is the **"Fair Share Distribution System Theory".**

Then I give some thoughts to the future and analyze: some possibilities of the relations, between different intelligent existences in the **"Interactions of Intelligent Species Theory",** a new way to conclude from our existing knowledge and experiences.

Part 1 - Chapter 1
Basics

Basics (cont.)

<u>To complete my investigations:</u> and to conclude based on some of our own theories no "Philosophical System" can be complete without investigating, analyzing and pointing into the direction of the "Future of Humanity". I try to do that to find solutions that build on our own theories the philosophy of McKaneism.

<u>Then it will be the time to:</u> organize and compare the existing "Schools of Thoughts" that existed before McKaneism.

I will do this in Part 1 for my "Beginners" and Part 2 for my "Advanced" readers. Don't you ever think that for the "Beginners" here, you don't need to think hard to follow the explanations and reasoning. My "Beginners" are not average, ordinary thinkers on this planet.

In the last chapter of Part 1, at the end of our journey of Part 1, we will revisit the scope of philosophy, the place of philosophy in our life and set the agenda for **a free philosophy and philosophical institution idea**, that does not exist in our 21st century yet. (Try to tell someone in China that you don't like their system, and let us see which re-educational camp you wake up in the next morning. They are an example here, but they are not the only ones.)

Now we are ready to move on to Chapter 2 and start analyzing what I define as our Fundamentals.

Part 1 - Chapter 2
Fundamentals, Matter vs. Mind, Space, Time, Energy, Directional Theory

Fundamentals

Most thought systems start with some fundamental question to say: Where to begin? If one thinks about building a "system", (I use "system" as a definition here in a very broad way) one needs to define the building blocks to build it from. For example: When someone builds a house, they need to use a variety of building blocks, such as lumber, cement, bricks etc., to complete the process. There, a finished house is a "system". When it comes to a "thought system" the "system" is built from thoughts that is defined by words. Also, the meaning of those words are very seriously defined so to make sure their meanings are the same for all the individuals who know these words. As one can see we jumped way ahead of us.   Why am I saying that?

Without extensively analyzing this subject now (we will do that in a later chapter) we can see how complicated this process gets, when one realizes we have a few hundred natural and a few dozen man-made languages. Some words don't have the same meaning or any meaning at all in another language. Sometimes we need to use phrases and complex sentences to translate from one language to another reflecting the same meaning of the thoughts. My choice is English for my publication to build my "thought system". English (American) is a second language for me. However, I practiced about half of my life "Hungarian" the other half "English" as my daily life language. To this point I think I am "equally bad" in both languages. However, because English is the most current one, that is the one I will try to use. (As I said previously we will spend a lot of time analyzing languages later.) Later we will revisit this question: Are we thinking language independent or within our own spoken language?

We don't answer this question now, we need to talk about only the framework of this publication. I intended to publish the first edition with only light editing, so any one can read my thoughts as they formed in my mind. That will create a great deal of authenticity and also a great deal of difficulty to read through the subject matters — its not easy by themselves any way or form at all in the first place and then the added challenge by my language imperfections.

Part 1 - Chapter 2
Fundamentals, Matter vs. Mind, Space, Time, Energy, Directional Theory

Fundamentals (cont.), Matter vs. Mind

In the future I am thinking to publish a (seriously edited by experts) second edition about five to ten years down the road, of course if I will find a reasonable amount of interest for it.

McKaneism, as a philosophy, tries to be observations and realistic logical conclusions based, so we will start from a few basic thoughts as our building blocks, then go for a more complex analysis. By the way, most other thought systems start from almost the same base.

The major questions to answer are:

What is the relation of the Matter vs. Mind?

What is the definition of Space?

What is the definition of Time?

Two new questions that I think are important:

What is the definition of Energy?

What is the Directional Theory?

By answering these questions we will create the fundamentals for McKaneism

Matter vs. Mind

What is the relation of the Matter vs. Mind?

All the existing "thought systems" try to answer this question one way or the other. The systems that answered "matter defines the mind" are called materialistic systems. The systems that said "mind over matter" are called idealistic systems. Also, we had some thoughts in the human history that said: I don't want to answer that question or simply stayed away from analyzing or answering any questions that try to answer this one or lead near to it. For example Marxism is a materialistic thought system, most religious thought systems are idealistic systems like Christianity, Islam, Judaism. For example Confucius in his thought system tried to avoid answering this question at all. This grouping of "thought systems" are historical and not my invention. Some of the first written notes that I know take us back to two, two and a half thousand years into the Greek existence.

Part 1 - Chapter 2
Fundamentals, Matter vs. Mind, Space, Time, Energy, Directional Theory

Matter vs. Mind (cont.)

The first time "thought systems" were departed and defined by their nature was Aristotle vs. Plato. Aristotle was the first who was called a real materialistic vs. Plato a real idealistic philosopher. Not because they didn't have predecessors but they were (living in the same time Aristotle attended Plato's school, then formed his own school of thoughts later.) philosophers who departed and defined the human thought process for the two different schools of thoughts.

Why is it important to decide on this question?

It is simple. The answer to this question will be one of the foundations of the "thought system" that one builds. Just like one can build a "Log house" or a "Brick house" based on the basic building material being wood (logs) or bricks.

As one can see, one can end up with the different house based on the basic building material, the same way one can end up with a different "thought system" answering this basic question differently. This is why it is very important how we answer this question. Also, because it is a building block of many following thoughts, it will determine the directions of observations and conclusions that one can make based on the starting point. For example, Christianity believes that a superior being, God, created the universe and everything in it. If we follow this thought process that will leave very little or no room at all for Darwinism, which strongly supports a theory of evolution based on the laws of nature. If one takes creation, strictly based on the Bible, that will leave very little room for the science of anthropology that tries to reveal the growth and evolving process of humanity through millions of years.

Here, I don't try to prove or disprove one or the other, simply try to show some examples that the decision on basic questions can take anyone to a different road to travel. The other reason why humanity tries to answer this question so hard is because we want to have a base, a starting point to start with and help to understand our origin. This will lead to a very complex thought process that the world around us is a static, well constructed, beginning to end, well designed existence, or a relative constantly changing environment with no beginning or end just continuous changes in the timeline.

Part 1 - Chapter 2
Fundamentals, Matter vs. Mind, Space, Time, Energy, Directional Theory

Matter vs. Mind (cont.)

As one can see, a search is on and will continue for a while, but now it is time to visit McKaneism and the thought process that this philosophy thinks about some of these basic questions of humanity. McKaneism's answer is very simple and far reaching for the next chapters for the Matter vs. Mind question:

This is an open question that we are (Humanity) not able to answer as of the year 2003 and before, and anyone who tries to give a defining answer will give a wrong answer!

Am I saying that all those great people ahead of me were wrong answering this question? The answer is: Yes, they all were wrong at their reasoning! One can ask me: What gives me a level of confidence to denounce all my predecessors who try to answer this very important question? My answer is to follow me on the historical facts, and a little logical reasoning, and you will end up with the same conclusion that I did!

One fact we know for sure, we didn't reach out to the end of our universe yet. We have no idea, (by proof) that there is any other intelligent life out there. (I am aware of UFO's but I didn't see yet an interview with someone out of our world!) It makes some sense to think that in this big universe there is a great possibility for other existence, but we have no scientific evidence to prove it. If we look at ourselves (based on some good scientific observation) it looks like we need the matter of mass of our brain to have the conscious working mind. So we observe that mind works on matter as a primary existence. (matter over mind or matter first.) However, we also observe the creative power of the mind. We know that a bridge can be created and exists in the mind before we build it. Then someone can conclude that the bridge is built from a matter of the molecules that stores the image in our mind. Then we can encounter some unexplained events when people think the human soul can separate from the body and live on

We have some ghost experiences, (some of them measured by instruments) that tell us that something is out there. And we can go on and go on and go on…

The final fact is, if anyone tries to be realistic and logical about it, we are not able to answer this question.

Part 1 - Chapter 2
Fundamentals, Matter vs. Mind, Space, Time, Energy, Directional Theory

Matter vs. Mind (cont.)

McKaneism is a Philosophy without answering this question or taking a side on the traditional way. More than that, as we will see later in more detail, it does not answer this question in the traditional way (nor precisely saying we don't know) and we will open a lot of room from many aspects of my Philosophy without limiting it. My conclusion is that searching for, or building a "Philosophy" to be "The Foundation of Humanity", I can't accept theories or conclusions that do not stand the strong base of facts and observations of those facts. In other words, we never have encountered the higher creating power so far, nor have we a proof of non-existence of it. It looks like the materialistic explanation makes more sense, but we are lacking the knowledge to rule out the other side.

In summary, humanity is only in the very beginning of it's journey of exploration of the many wonders of the universe. Until we find conclusive evidence through observation and facts for this question (not through theoretical reasoning that all current and previous existing thought processes do or did!) we will leave it open and say this question can't be answered based on the knowledge of humanity in the 21st century (year 2003)! Finally we have a first building block of McKaneism, the conclusion of Matter vs. Mind question. The conclusion is that we observe matter and we observe mind, so both exist in reality — but humanity based on the 21st century knowledge does not know the answer for the question: Which one is over the other?

Mind over matter or matter over mind? McKaneism says: We don't know! What is the relation of the Matter vs. Mind? They both do exist ! We can give a definition for both because both exist, we can observe that. So here are my definitions:

Matter is an objective existence! When we see a rock, building, animal, human, etc., the common thing in those objects are that they all exist, the difference is their form of appearance.

Mind is a self and surrounding's realizing entity! We know through our senses we can observe and realize our surroundings and ourselves. We can say our mind is reflecting like a mirror the material reality around us.

24

Part 1 - Chapter 2
Fundamentals, Matter vs. Mind, Space, Time, Energy, Directional Theory

Matter vs. Mind (cont.)

We know this much for sure and this will be my conclusion of the Matter vs. Mind for now, so we can start building our familiar chart with the first defining base.

Chart 3

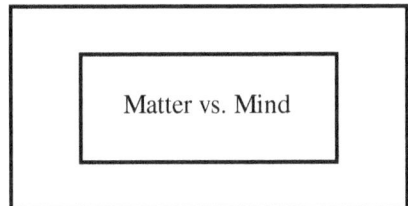

Space

The next element in the foundations of McKaneism (Foundation of Humanity) is Space. Most "thought systems" don't pay much attention to Space, because it seems self evident. If you think about it seriously the definition is not as simple as it looks like.

Especially if one starts learning the physics of space one can realize it isn't simple at all. For now I will define space as: **Space** is an environment where all the matters and minds exist and functioning. For easy understanding, with an example, Space is a holding area like the space of the Solar System that according to our knowledge extends about one light year from the center of the Sun. If one can imagine a basketball with the radius of one light year, that is the gravitational force of the sun, or in other words our Solar System, Our extended home.

Of course this is very virtual for most of us because very few people in life can imagine the distance that a light beam can travel within a year. It is the speed that the light can travel around the equator of Earth about 7.5 times a second. Imagine that speed of light traveling for a whole year. It is 365 days times 24 hours times 60 minutes times 60 seconds times 7.5 times around equator. It is a real "great distance". That is the gravitational force of our Sun. It is too imaginary for most of us because no human mind had real observation experience to comprehend that space and distance yet. Now it is time to add Space to our chart and move on to the next basic element of fundamentals of McKaneism

Part 1 - Chapter 2
Fundamentals, Matter vs. Mind, Space, Time, Energy, Directional Theory

Space (cont.)

<u>Chart 4</u>

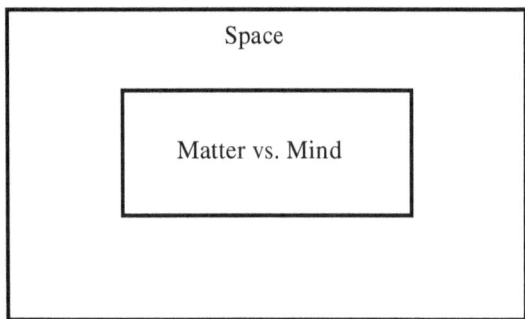

Time

The next element in the foundations of McKaneism (The Foundation of Humanity) is Time. We can start with the same thought of definition as we did defining Space. Most "thought systems" don't pay much attention to Time because it seems self evident. If you think about it seriously the definition is not as simple as it looks. When we think about time in the daily life we think about clocks and watches. However, those are only instruments to measure time. They are not a definition of time at all. For now I will define time as: **Time** is a dynamic event between two static points. (We will revisit this in a later chapter!) For example: When I drive from my home to the my workplace in the distance is Space, the (measurable) "dynamic event" here is the driving observed as Time.

Time is very close to movement which is very close to Energy, which will be our next basic element. Without movement Time does not exist as we know it today. For example, if you sit down in your living room and fix your eyes on one of your pictures on the wall and sit without movement, you will not be able to tell if you were sitting there five minutes or one hour. In this example, you won't be able to observe any events, so for you nothing dynamic exists at this moment. However, if you are (in the same example) from time to time able look up to the clock on the wall, the movements of the hands on the clock (like events) will tell you the changes, so you can observe the events between two static points. (the hands of the clock move from five to six, it can be five minutes or an hour to pass by.)

Part 1 - Chapter 2
Fundamentals, Matter vs. Mind, Space, Time, Energy, Directional Theory

Time (cont.)

We will revisit Time as I promised later in other chapters analyzing logic and logical events. We will see how Time becomes an important factor when we differentiate between formal logic vs. dialectic logic.

For now let us add Time to our chart and move on to the next basic element of fundamentals of McKaneism, Energy.

Chart 5

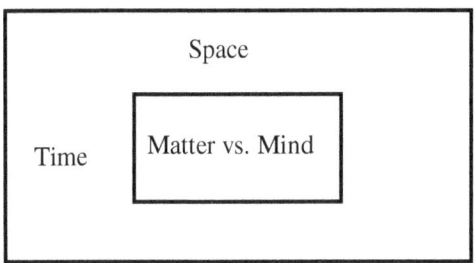

Energy

Energy, and movement, are the next fundamentals in McKaneism. Mostly the materialistic "thought systems" pay attention to movements and energies. It is an attribute to the matter. We can't analyze matter without movements, and can't experience movement without energy existence. Most of us know from early school lessons of science (physics) that the atom exists as an electron circle around a nucleus (or center) of that atom. (We use this model of representation of the atoms for an easier example — in the serious nuclear physics it is far more complex than we even can imagine!) So here in our example we see that where atoms exist the movements of those atoms with their own energy levels are also present.

Even though a table looks like a solid wood structure, if we extend our sense of vision to the atomic level, we will see that the table inside is a lot of atoms with a lot of movements. It was always fascinating for me to imagine that the solid objects are in constant movements of atoms inside.

Part 1 - Chapter 2
Fundamentals, Matter vs. Mind, Space, Time, Energy, Directional Theory

Energy (cont.)

We can conclude based on our knowledge of physics that no matter can exist without the movement of atoms that translate to a given energy level. I don't know any form of matter or existence that can exist with zero energy or without the movement of the building blocks in our example, the atoms. All our observations in our daily life provide countless examples for Energy transfer from one form to the other. For example: The energy in our food that we consume transforms into energy that moves our muscles. The electric energy heats up the soup we are cooking and the list can go on and on. One of the basic principles of physics is that energy won't get created from nothing or does not disappear to nothing, only constantly changing from one form to the other. I realize that these are not the perfect scientific definitions, but I think they are properly representing the thoughts that I am trying to communicate, through some of these examples. I will revisit Energy in a later chapter again and I will spend more time analyzing it. For now I will define Energy as: **Energy** is an attribute of matter and mind and undividable from matter or mind regardless of the appearance and form of them. For now let us add Energy to our chart and move on to the next basic element of fundamentals of McKaneism, "The Directional Theory".

Chart 6

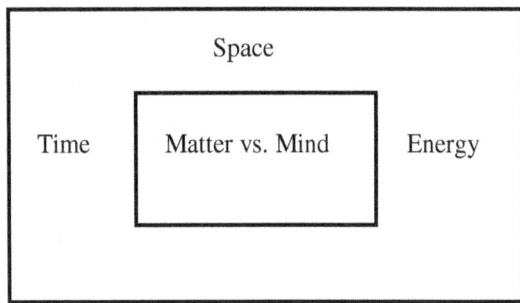

The Directional Theory

The Directional Theory is new and specific to McKaneism. (As far as I know!) In this form of definition I didn't find this thought process in any other previous "Thought Systems".

Part 1 - Chapter 2
Fundamentals, Matter vs. Mind, Space, Time, Energy, Directional Theory

The Directional Theory (cont.)

However, part of the theory already exists, so I didn't invent or create this theory, only extended some of the existing ones building on some of my predecessors. (As you will see through out this book that is another rule of Humanity, that one generations' knowledge is built on the previous one.) Two very important places where one can find "part" of "The Directional Theory". The first comes from biology, from Darwin's Evolution Theory. He teaches that given an extended period of time the biological species (including humans) evolve or grow: from the simple to the complex form, based on the influences of the environment. If we look at the millions of years that he must analyze in his theory we can probably have endless arguments. However, if one observes a growing process of his or her own child, one can see how the theory stands solid in the small dimension, where one kid grows into an adult thereby changing, growing, evolving based on the influence of the environment.

Interestingly we find generations of doctors, actors, musicians, etc. from parents to children through many generations. Looking at their environment, upbringing, influences, peer-pressures, one can observe that more than coincidence exists in their environments. Unfortunately we can observe the negative side of the coin when we analyze the depressed neighborhoods in our USA, one can find sad conclusions that those environments are the negative influences of many youngsters, whose lives take a turn many times in the wrong direction.

Like always, these are examples to highlight my subject matter and to make communication easier, so at this point I don't get into any serious arguments with anyone about the why. The second place where "part" of "The Directional Theory" comes from is Marxism. Unfortunately, because Marxism has been equated with Communism throughout history, the word Marxism looked upon as evil in the western societies. We will analyze this a lot more in the chapter that analyzes Sociology. For now one needs to understand that Marxism is a "thought process" just like Christianity or Islam or Judaism, and many others, but only in the sense as a definition they are all "thought processes". For us here the interesting pieces from the thoughts of Marx, the following two special thoughts.

Part 1 - Chapter 2
Fundamentals, Matter vs. Mind, Space, Time, Energy, Directional Theory

The Directional Theory (cont.)

Marx defined that if something grows as a **quantity** into a direction, at a certain point we experience a different **quality**. I won't extend more on this discovery now, and it does not matter that one believes in it or not. I will just provide some observable examples, how much it is truth, in our society of what he was saying, that we even use different words to describe the quantity-quality changes. If we look at our kids again, they grow in (quantity) weight, etc. and we give them a different name (quality) from baby to toddler to kid to teenager to young adult. The other observation coming from Lenin, who studying history realized that occasionally the quantity-quality changes are happening extremely fast that he called it revolutions. We all know this for a fact, that throughout human history, social forms are changing through a very fast and sometimes violent process. (EG: French Revolution, USA Independence.) The Directional Theory of McKaneism is the continuation, extension, generalizations of these thought processes.

So the "part" is that they were addressing the **slow** quantity to quality transformation of the biological life forms and the **fast** revolutionary change of society for a different quality. **My Directional Theory goes a lot further**. The quantity-quality changes can be slow or fast and also can go ahead or in the reverse direction. More then that the direction can be any direction. We will revisit these thoughts in the later chapters in more detail. For now the definition of **The Directional Theory** is: The matter or mind based on the level of energy can move from simple to complex or complex to simple, or any other way, either slow or fast or any speed. (I will return to these thoughts in Part 2.)

Before we insert the final block to our basic chart let us view a few examples that anyone can observe in real life. For example: We all can observe any living form to grow from small to full grown, that is a **simple to more complex slow** process in life. Unfortunately, at some point in our life we will experience the passing of one of our loved relatives, that is the **slow reversed process of the complex mind and body to more simple** direction. It sounds like a "heartless" example, but as we will see, the processes of "The Directional Theory" care very little or not at all for our human emotional feelings about our life.

Part 1 - Chapter 2
Fundamentals, Matter vs. Mind, Space, Time, Energy, Directional Theory

The Directional Theory (cont.)

An example of a **complex to simple fast directional process** is the eruptions of a volcano. It will flow through, and perish all existing life on it's way, sometimes within a matter of hours. The same process for a forest or brush fire. Finally, an example for **the simple to complex fast event** in "The Directional Theory" is the activity of our Sun. A more simple hydrogen turning into a more complex helium in every fraction of a second.

I am sure anyone with a little imagination can come up with countless examples in our every day life. It isn't a surprise, it is a factual discovery that "The Directional Theory" is a very viable part of our existence, and you will see many serious implications as we progress further. For the best example — for any direction in my theory is the river. The running water can take on any direction depending on only the quantity of the water in the river. As we all experience, unfortunately, sometimes the great quantities of water can create a very troublesome river for us leaving the original river flow and creating an any direction, different quality river. One more thought for those of you who will think this is a Communist book, because of mentioned names, (Marx, etc.) I have interesting news for you. Even in the "thought process" of the Bible, when God creates the universe, he creates something (universe) starting from nothing (nonexistence of universe), and he did it in six days. So his directional process in this case is a simple to complex fast directional process! It is really amazing how "The Directional Theory" of McKaneism can stand on this very solid ground.

Now let us add the Directional Theory to our chart completing it, and spend just a little more time to sum up the basic fundamentals of McKaneism.

Chart 7

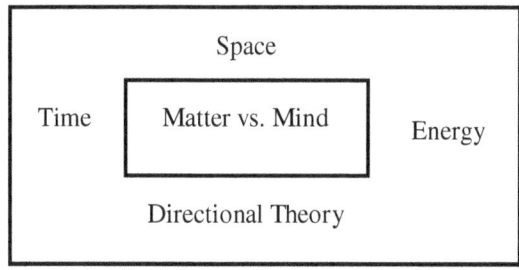

Part 1 - Chapter 2
Fundamentals, Matter vs. Mind, Space, Time, Energy, Directional Theory

The Directional Theory (cont.)

Now we can see in Chart 7 (in the picture worth a thousand word context) the basic Fundamentals of "The Foundation of Humanity" (McKaneism). When we continue progressing further in the following chapters one can realize how simple and easy it is to understand and build a more and more complex "thought system" based on these fundamentals. We can see so far that **Matter** and/or **Mind** based on the **Energy** level in them constantly moving into a direction based on **The Directional Theory** in **Space** and **Time**.

It looks like a straight forward, very logical argument, easy to see and understand, to anyone who is willing to think on their own, instead of taking someone's word for anything. Like I did a few times before, I try to remind you that the best way to read this book is to think these thoughts over, find some examples on your own, and draw your own conclusions. Now we can move on and see how we can build a more complex thought system on these basic fundamentals.

Part 1 - Chapter 3
Mechanics, Physics, Chemistry

Mechanics

The next subject to move on to is **Mechanics**. Mechanics is a subset of Physics, and for an easy head start, we choose Mechanics as our next place to observe and investigate. As some of you may remember, when we concluded about Matter vs. Mind, we realized that they may not exist together all the time. For example, we know that we have existing Matter on Mars — as a matter of fact, the whole planet has different forms of matter, however, we have no evidence of an existing Mind on Mars, as of today. To build a realistic philosophy one must find a step by step, easy to follow, logical approach to succeed. Mechanics is a very important, realistic science that provides many great advantages for humankind. Many times (maybe too many times) I have to remind my readers that this is a philosophy publication (intended to be!) so we don't jump into a full description of mechanics as a subset of physics, because this small volume would be very insufficient to describe even a small part of mechanics. However, we will do two major things here.

One, to put mechanics in a place where it logically belongs, and the second, we will look at some of the attributes of mechanics, from a philosophical point of view. The most simple movements of matter are being analyzed in Mechanics (However, I am not saying that Mechanics itself is simple at all). But, a few comments before we go further. Each time we approach a science in this publication we focus on the science's place in philosophy, and it's interaction with philosophy. Anyone who wants to check the correctness of my statements must go on to read more on those subjects and sciences if one missed acquiring the knowledge in one's school years.

With the single word "subset" I am referring to a part of mathematics, called the Set Theory (and maybe called many other names also). This theory itself probably has a larger volume of publication than this book. For example, the principles of mechanics help us build houses, cars, roads, airplanes, bridges, smaller items like washing machines and toasters, and we can go on and on. You can imagine that if one needed a blueprint to do something, the principles of mechanics would be involved. Mechanics is one of the most difficult subject matters for an engineering student in our colleges and universities.

Part 1 - Chapter 3
Mechanics, Physics, Chemistry

Mechanics (cont.)

So, for our investigation we will try to summarize the main and philosophically important attributes of mechanics.

It deals with matter, its energy level and movements that are "slow". We will explain "slow" in a moment. For an easy to understand example, I will choose a pool table. Many of us have had a good time in our lives playing pool, so I thought it would be an easier example for people to understand. If we place a ball on the pool table we all know that it will stay there until we push it to the desired direction. So this ball will behave as a **Matter**, using the **Energy** that we delivered through the push, and it will move on the table to a **Direction** in **Space** and **Time.** Interestingly, we find our **Fundamentals** perfectly fit within the realm of **Mechanics.** Of course, it is not a surprise because they are a "subset" of mechanics. Another thing we can observe here is that the ball on the pool table is rolling within a relative "slow" speed. This is another attribute of mechanics, that is to say mechanics investigates objects moving with a slower speed. We see some "Relativity" here, but I won't address it now, it will come later.

I will look at this one example to get some feel about this "Relativity" that I am talking about. When an airplane moves 600 miles per hour, it is a fast speed in our every day life. But it can't compare with the speed of light. The "slow" movements in mechanics are relatively fast compared to the capability of humans (e.g. walking 4 miles/hours), but they are "slow" compared to the speed of light. We will move into that extension when we move on to investigate Physics. For now let us spend a little more time around mechanics. When we defined **Space** as one of our fundamentals, on purpose we didn't go into analyzing dimensions. Now we will venture into that area a little. On the pool table we move each ball on a flat two dimensional surface. In our practical life we can get by as humanity in the three dimensional world.

We can understand our Solar system as such objects like Earth, Mars etc. moving around a Sun with "slow" speed in three dimensional space.

Part 1 - Chapter 3
Mechanics, Physics, Chemistry

Mechanics (cont.)

I think from these examples we can come closer to understanding that from our stand point of philosophy, **Mechanics** is a science that focuses on the movements of objects and their energy levels. If we go back to one of our examples, we can extend on this.

When we observe a human being walking we can say that they are **Matter** and **Mind** moving with "slow" speed, to a **Direction** using its **Energy** in **Space** and **Time.**

Again we find our **Fundamentals** perfectly fitting within **Mechanics.** It is important to place mechanics as a science in the right place in our "thought system".

Let us update our chart adding **Mechanics** to it. It may not look like an important step or discovery at this time but we will see later, as we keep building our system, that the science of mechanics is very important for every living species — and later, we will insert into our system a chapter on Biology.

For now, here is our current and updated thought system of McKaneism for Humankind:

Chart 8

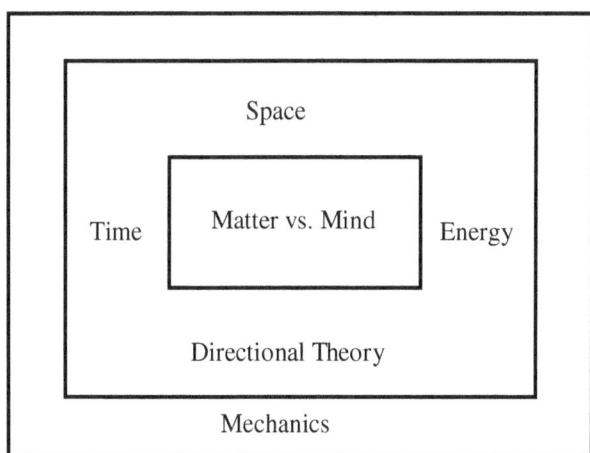

Now we can move on to the next subject matter of our investigation, and that is Physics. Another science that is interesting, but I hope not too surprising for anybody. We will move on and will think about things that are moving "fast" instead of "slow", but as you guessed correctly, they include all the characteristics of the "slow" moving ones and more.

Part 1 - Chapter 3
Mechanics, Physics, Chemistry

Physics

The next subject in our journey is Physics. Physics is a science that includes all the knowledge of mechanics and more. For example, the science of hydrology, astrology, nuclear physics, and more.

It is important that we extend our thought system from the object or matter moving slow in space to objects traveling with the speed of light in space. For example, the electrons around the nucleus of the atoms or light itself. In this world of physics we need very sophisticated instruments to extend our capability to observe, because most of the events on the level of nuclear science are not observable with our senses.

Here we are introducing an interesting way of discovering the knowledge of physics that was actually introduced by Einstein. We can call these "Thought Experiments", because all the physical experiments are conducted in our mind. (Actually in Einstein's mind.) When we face the speed of light, our sensors are not helping our observations. It is very difficult to imagine, but when one turns on a flashlight and points it to the sky, the light will travel away from the flashlight at the speed of light into space. It is even more difficult to imagine for an observer, but if you had two trains, going side by side at the speed of light (and investigate from the stand point of physics), what will happen if you flash your flashlight from one train window to the other? This is the fascinating world of Einstein, who spent countless hours in his physical discoveries, to investigate these situations within his "thought experiments".

We can see that our investigation moved from the "slow" moving, easier to observe objects, to the very small (in nuclear physics) or the very large (in astronomy) objects, which some of them are "fast" moving (as fast as speed of light). The interesting observations in our philosophy are the following: We can have systems, like our Solar system, where the rules and principles of physics exist without an exploring or observing Mind present. We can observe a "system of matters" functioning by the rule of physics.

To bring back one of our previous examples, with the rules of physics we can look inside our rolling ball on the pool table, using our advanced instruments and mathematics, and discover the wonders of the physical laws inside the "simple" eight ball.

Part 1 - Chapter 3
Mechanics, Physics, Chemistry

Physics (cont.)

Another example, we can imagine a space craft that starts moving at 600 miles per hour (like an airplane) then starts accelerating to the speed of light (I know it is impossible for today's science, so do this as a "thought experiment").

At the beginning, whatever happens on this craft can be described and explained by the laws of mechanics, however, when we get closer and closer to the speed of light, we need to start using a theory of relativity to explain the events on this craft.

In summary, I just tried to highlight the relations between the science of mechanics and the science of physics. I tried to find some examples (I hope I did) that give some view of my statement that the science of mechanics is a subset of the science of physics. Here in the world of physics we also can find our basic fundamentals — matter, space, time. I hope that my readers are starting to realize that I am building a "Thought System" or "Thought Process" where "The Directional Theory" also strongly applies.

We are looking at this in a way that so far, where we are going, a "simple" to a "more complex" set of rules and principles of the sciences that analyze, investigate, and try to discover all aspects, rules, and principles of matter, space, time and energy, are appearing (Our "good old" re-appearing Fundamentals!).

I have to repeat myself, when I say "simple", I am not stating that the thousands of public library books on mechanics is a "simple" thing. I am using "simple" as a relative term realizing that compared to the even thousands of more books on physics, these books are "more complex" and more voluminous than the books on mechanics.

The main reason why things get more and more complex is because as we move ahead in our book, organizing the sciences in my "Thought System", we will discover that every next step leads us to a more complex form of Matter (and/or Mind) with more complex rules to describe them.

Like always, before we move to the next subject, we need to say a few more words about the "simple" to "complex" observation to make sure that we have the same view on this before we move on deeper into to our analysis.

Part 1 - Chapter 3
Mechanics, Physics, Chemistry

Physics (cont.)

The analysis in **mechanics**, if we think about our pool table, was focused around the fact that we apply a force on the ball and the ball will roll with a certain speed, bouncing from a bumper of the pool table and ending up somewhere within the well defined boundaries of the pool table. When we moved on to **physics**, our interest of observations were getting more in-depth.

We wanted to see "inside" the ball and then we asked the question "what happens if the pool table is the size of the universe and we pushed the ball at the speed of light?" This question brought more complex investigation into our process.

Let us place physics on our chart and then we can move on to the last subject of this chapter, Chemistry.

Chart 9

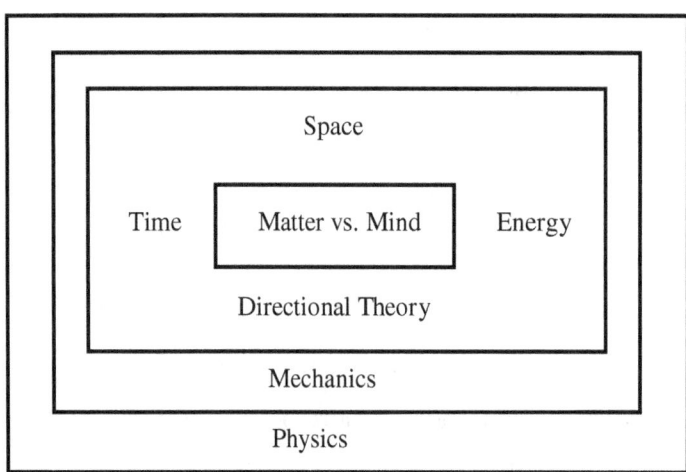

Finally, in this chapter we moved on to ask the question "how are all these matters around us built?", "What kind of construction are they?" We find the answers for these questions in the world of **chemistry**.

The analysis of hydrological pressure in physics includes the observation of water. We know that if I am interested in having a shower built in my house, I am interested in bringing the appropriate water pressure to that shower.

Part 1 - Chapter 3
Mechanics, Physics, Chemistry

Chemistry

I am staying within the boundaries of physics to calculate all the numbers to accomplish my task. I know that I will get H_2O molecules in my shower (sounds very stupid saying that someone thinks about this while taking a shower) but I don't think I care as long as the water is coming with good pressure. The story changes very seriously if I decide that I want some clean drinking water in my house.

In this case, I am not interested that the one electron circles with the speed of light around the one proton in the hydrogen atom, (that is the subject of physics) I am more interested in what other kind of matter may be hanging around in my drinking water.

That is my move to the direction of Chemistry. Chemistry is the science that will analyze the compounds in my water and with the available tests make sure that I have clean drinking water. One can see that the physical attributes of the H_2O won't change, so based on this knowledge, I can conclude that Chemistry will incorporate all the previous rules of physics as a subset of Chemistry. Just to put the "Set Theory" to rest, for those of you who were never interested in mathematics on that level, "Subset" means "includes" in the daily language

I think most of you know that toy set, that initiated from Russia originally, when one small wooden toy doll fits into another bigger wooden toy doll, then another bigger, and so on. So, holding the biggest toy doll and then opening it, you find a smaller one in it, then another smaller one, and on and on. Another good example is those kid cube toys where on one side of the cube it is open and one smaller cube fits in the bigger cube, and another bigger cube, and on and on. In short, the set theory for one set includes all of the next work, as these examples do with kids toys. The most important aspect for us here is that each time I am using the subset definition, I am stating that the "Set", like Chemistry, includes Physics as a "Subset" with all the rules and principles of Physics.

Chemistry is basically interested in how matter builds from different kinds of atoms. Earlier in history, some of the scientists of chemistry discovered the "periodic system" of elements. Now we know that different atoms exist and they form compounds, mixing with each other.

Part 1 - Chapter 3
Mechanics, Physics, Chemistry

Chemistry (cont.)

We divide chemistry into two major investigational areas, one is the non-carbon based chemistry, the other is the carbon based chemistry. We (humanity) discovered that we have one element, the carbon ("C"), in our "periodic system" which is capable of forming a special, interesting, and looks like endless variety of compounds that hold a great importance for all of us. This is the element that plays a very important role of the formation of life. (That will be our next chapter, to investigate under Biology.)

For now, let us place Chemistry in the place that it belongs, the study of the rules and principles of elements and compounds and the way they form and break and reform again.

One can see that our **Foundations** are still solid because Chemistry investigates the **Matter** that exists in **Space** and **Time**, and follows the rules of **The Directional Theory**. Let us incorporate Chemistry into our chart so we can see "The Foundation of Humanity" as it develops in front of our own eyes.

Chart 10

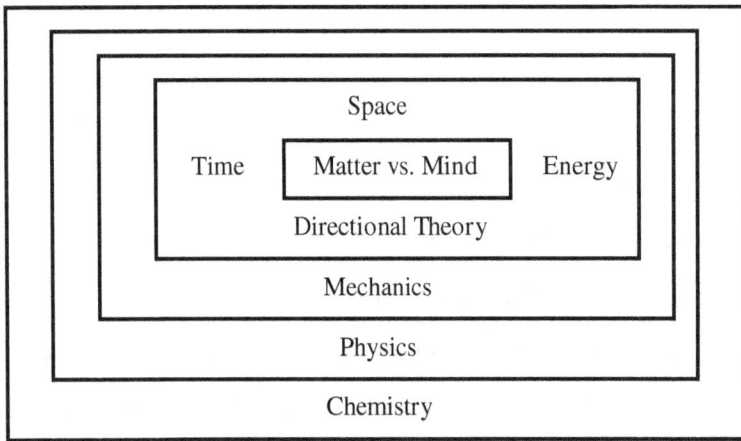

We are just about finished with Chapter 3, and as anyone can see, we are building a "Thought System" or "Philosophy" or we can call it McKaneism here.

So far, we put down the basic foundations and a few sciences that are built on those foundations. At this point we can stop for a minute and ask anyone who trying to challenge this philosophy to do just that.

Part 1 - Chapter 3
Mechanics, Physics, Chemistry

Chemistry (cont.)

All we said so far is based on the knowledge of humanity and some light realistic thinking; anyone with a basic school education can follow the thought process and can make his or her own conclusions. In the previous chapters, I am not intending to be complicated or difficult to my readers. I try to build an easy to follow **logical, reality reflecting system** that makes sense. This approach will continue into the next chapters, however, we will be moving into a more interesting, more debatable, and as we know, a more complex subject matter.

After Chemistry we will investigate Biology where we encounter the appearance of life that will change our discussion from a non-living to a living entity. It is a good time to take a break for a few minutes after finishing the reading of Chapter 3, with the understanding that the best is yet to come.

Part 1 - Chapter 4
Biology, Attributes of Living Entities Theory

Biology

Our investigations, until now, were confined to subject matters that were observed by our mind, but that also exist without any sign of the mind at all. Up to and through the level of Chemistry, the matter exists, interferes, and changes form and shape, from one to the other, based on the energy that influences those changes. When I say "up to and through" I am inferring that in this "Thought Process" we view things from the simple to more complex formation, where the more complex includes all the attributes of the more simple. For example, Physics as a science, includes Mechanics as a science.

These observations form the "thought process" of most materialistic philosophy that points out that we can easily imagine, or if we want, we can observe a "Solar system" where all the rules of mechanics, physics, chemistry apply and working, yet no living entity exists. (Other than Earth of course.) It makes a strong argument also that matter can exist without mind, but looks like mind can't exist without the matter, because all the minds we observed up to today are based and built on matter. As I made the statement earlier, it sounds logical, however, our very small knowledge of the huge universe does not provide proper evidence to really make that conclusion.

Moving higher in the complexity of matter, we encounter Biology. Biology is investigating the "Living" existences. In the daily language, living is an easy to understand term, but as many times before, it is not as easy to define the point of view philosophically. We will follow the popular definitions now that "living" existences show vital signs. In other words, they are more or less capable of understanding their environment; they interact and respond to that environment.

Here we discover the importance of the "Carbon based" chemicals, because these chemicals make it possible, through their flexibility, to connect to build large, complex living existences. So, the matter forms a "system" called a "neuron system" that has a special characteristic to identify and respond to the environment.

The level of responses are different based on the complexity of the "neuron system".

Part 1 - Chapter 4
Biology, Attributes of Living Entities Theory

Biology (cont.)

On the following few pages, I will try to analyze and organize some simple ways to introduce and group the great variety of existences that we encountered in our environment, planet Earth.

As always, our focus is philosophical and we will try to investigate from the view point of philosophy, and avoid the complex road of biological explanation because of the lack of volume in this book and a lack of knowledge of the author. But I know that as much as our tools of observation become more advanced, we encounter more and more life forms that we had no knowledge of before. To our surprise, we discovered that in a cubic foot of soil of the rainforest we find millions of living organisms, or we discovered living things under the deep water that we had never seen before. We discovered living organisms that can exist on the bottom of a volcanic seabed in 600° F temperature. I can go on and on with this list of wonders for many-many pages, but instead, let us look at Biology as a subject from the philosophical point of view of us. I will define the life signs that we encounter when we interact with life forms, and will investigate what kind of life signs need to exist to call any matter a living one. Then I will introduce a new summarization of thoughts that I call "Attributes of Living Entities Theory", that will analyze some basic capabilities and attributes of the living existences.

It will be interesting to discover, for example, that "Human Health" is not a right (as our Democrats think about it: the right for health care), or is not a commodity (as our Republicans think about it: you buy it through insurance just like you buy a piece of bread), it is an "attribute of a living existence" — in this case, Human Beings. This illustrates another example as to why philosophy is important, because understanding these three simple statements can help us design a health care system with an understanding of a great importance, that healthy individuals, with good healthcare, define a healthy society.

Moving on now, one of the most important aspects of biology is to study the existence of Matter and Mind in biological entities that enable it to respond to and adopt to the effects of the environment which those entities exist within.

Part 1 - Chapter 4
Biology, Attributes of Living Entities Theory

Biology (cont.)

Observing a piece of rock ("world of physics"), a rock is not capable to adopt to or respond to the environment. If some force (like we kick it) makes it move, it will move until the original energy provided is used up by the move. We observe the same in the flow of water ("world of chemistry"), that is it starts flowing from higher ground and ends up in a lower level. However, we can observe animals ("world of biology") that seek out shade in the hot sunny day to cool their body temperature or fish that are able to swim upstream to feed and satisfy their hunger. At this level of biology we observe something new that we did not find before, and it is a responding mind. Here we must ask two important questions:

How can we advance from a chemical to a biological existence?

Is evolution theory correct?

In the beginning of the history of humanity, we viewed biology as the study of plants and animals. During human existence, we used the word "medical science", that is, we tried to extend the observation and study of the human body and mind to fix or correct the broken things in it. I will revisit medical science in the next chapter when we analyze Psychology. Knowing that the human body and mind together form a biological unit, we then analyze biology from the point of philosophy.

Starting with the first important question. Yes, we are moving from chemistry to biology because the complexity of observations are increasing. The matter becomes more complex and we observe the function of the mind that we didn't before. The simple step from chemical to biological existence has been very highly debated in the past, and even the 21st century. The theory of evolution from Charles Darwin and his followers think it is a logical step to picture an environment where the more complex carbon based chemicals can turn into a single cell type of living form. That experiment can be performed in the laboratories, in an observed and controlled environment where we can observe and record every step of the process.

For example: When we are successful and create a single living one cell organism; Then : Are we giving proof for Darwin's theory of evolution, or are we giving proof to the creation theory where a higher level of existence creates a lower level one?

44

Part 1 - Chapter 4
Biology, Attributes of Living Entities Theory

Biology (cont.)

The answer would be the same as most of the world's religious theories since the beginning of our existence. Like we tried to explain in the chapter of Matter vs. Mind, we are back to the same problem. The only positive proof would be, one way or the other, is if we could travel back in time until we can find and record the beginning.

It is impossible for us now and we have no knowledge that it will be or won't be possible at all in the future. We can say without concluding on those important but hard to prove theories that we can view and observe biology, keeping our focus on the common, easy to observe attributes of many of the biological existences.

To call something living and make it part of the world of biology we have to observe certain reactions to the environment of Matter and Mind. These are the following:

1. **Self Preservation.**
2. **Energy Consumption.**
3. **Reproduction. (Continuation of the species)**
4. **Capability of movement in space and time.**
5. **The matter and mind responses to the environment.**

Based on our observations today, our science groups tell us of the living existences in three major groups: **Plant life forms, Animal life forms, Human life form.**

Today's science discovers every day something new and interesting from every group. Thousands of scientists observe and analyze the different aspects of living existences. As far as I know, we (humanity) didn't find an existing mind at the "plant life forms" area. I know some interesting observations, for example, that playing different kinds of music effect different levels of the growth process of some plant life. Is there any mind there that feels more comfortable or less disturbed in a certain environment? I don't know, but the observations are interesting

Looking into the Animal life forms, our investigations intensify. Maybe because they are more interesting as an existence or because some of them show some signs that we humans do. We definitely can tell that we are observing a responding mind in the animal world. For example, the search for food or the alert status against danger, etc.

Part 1 - Chapter 4
Biology, Attributes of Living Entities Theory

Biology (cont.)

We are spending a great deal of time, for obvious reasons, to observe, investigate and discover the wonders of the third group, that is the human existence. We analyze the biological, that is, the building of the human body and mind and also we try to discover the working signatures of our working mind — and that will lead us to the next level of complexity, that is Psychology, in the next chapter.

Let us distinguish, from the point of McKaneism, where we draw the line of observations between biology and psychology from the standpoint of our philosophy. We need to introduce here an important categorization of the living existences based on point 5. "The matter and mind responses to the environment". We can group all biological existences into three levels based on point 5.

Here they are:

1. **Level 1**, Existences with a simple **Reflex** response to the environment.

2. **Level 2**, Existences with a simple or complex **Emotional** response to the environment.

3. **Level 3**, Existences with simple or complex **Logical/Reasoning** response to the environment.

The complex: **Level 2 / Level 3** responses are what we will analyze in the next chapter, Psychology, but before we do that, let us continue with biology.

The following observation is very important, and I will refer to it many more times in this book. For the easier reference, I will create Chart 11 to reflect this observation.

Chart 11

Level 1, Existences with a simple **Reflex** response to the environment
Level 2, Existences with a simple or complex **Emotional** response to the environment
Level 3, Existences with simple or complex **Logical/Reasoning** response to the environment.

Part 1 - Chapter 4
Biology, Attributes of Living Entities Theory

Biology (cont.)

Here are a few examples for the three levels of responses. Every newborn baby knows the source of the food by its reflexes. These are genetically coded simple reflexes, Level 1 responses for self preservation. When we humans grow, we develop our complex emotional system, that is, we lead or develop to Level 2 responses, emotional responses like sadness or happiness. Then, we develop Level 3 responses like reasoning thinking, analyzing science and facts around us. These are just a few examples, but I am sure anyone can find many others in our living environment.

Just to point out, this is not a simple observation as it looks like in the world of biology. Think about the behavior of the herd of elephants when they find a bunch of elephant bones. They spend a great amount of time analyzing those. At this point I just have to ask the questions: Are they thinking at Level 3?, or are they feeling emotions at Level 2?, or do they just have a reflex reaction, that of Level 1?

 The answer does not look simple for me. It takes a great amount of knowledge of biology to answer those questions. And I don't think we (Science of Humanity) really have an answer yet. It is not too important to answer from our standpoint; we can continue with our philosophical thought process without those answers.

The next interesting observation of mine is that biological existences are capable of reacting to the environment because they have "senses". They have the capability to sense the energy level changes or energy volumes coming from their environment.

These senses are:

1. The sensors for **Seeing**.

2. The sensors for **Hearing**.

3. The sensors for **Smelling**.

4. The sensors for **Tasting**.

5. The sensors for **Touching**.

6. The **6th sense**?   (Electromagnetic, Gravitational fields changes around the existence?!)

Part 1 - Chapter 4
Biology, Attributes of Living Entities Theory

Biology (cont.), Attributes of Living Entities Theory

Not every existing, living thing has all those senses. Some of them have some more refined senses and some are lacking others. I think we (healthy humans) have all of these senses. As far as the 6th sense goes, we may be joking about it a lot some time, yet I have a few interesting observations.

I observed a nature show from Africa, where the lions and zebras were walking close by and they didn't pay any attention to each other at all. How is it that the zebra knows that it isn't a menu item this time?

Another picture I observed, where divers dive between sharks and the sharks just approaching them with curiosity, but they don't attack. One of the divers believed that if he thought and behaved with no fear, then the sharks are somehow sensing it.

From our human environment, the Chinese believed for many thousands of years that through the human body a mysterious energy, the "Chi", is flowing. I observed Chinese medical charts of humans (used in acupuncture) where they show lines around the human body showing energy flow the same way as we outline the electro-magnetic field around our planet Earth. Are these signs of the existence of the 6th sense? I don't know the answer but these observations are worth noticing. We performed some observations so far from the area of biology.

Attributes of Living Entities Theory

Now we can take an important next step in analyzing, or I can say introduce something new, that I call in my philosophy the "Attributes of Living Entities Theory". It is a very important part of my philosophy because the understanding of this theory will lead us in the next chapter to extend it to the "Attributes of Human Beings Theory", that will then extend our understanding, as far as human behaviors go, on the social level in a society.

For example: when one sees a small child having fun torturing pets, like cats and dogs, one should pay attention, because maybe it is a sign of a genetic problem that can lead to a disturbed adult, who sees no problem in torturing other adults.

Part 1 - Chapter 4
Biology, Attributes of Living Entities Theory

Attributes of Living Entities Theory (cont.)

But before we get ahead of ourselves, let us stay within the "Attributes of Living Entities Theory" in the world of biology. Each biological entity (and that is my observation) has their own characteristic or a set of attributes that define their existence.

Each of them have **General Attributes**, that is attributes like the awareness of **Self Preservation** and **Energy Consumption** (in other words, drinking water and eating food they can turn into other energy, for example muscle energy for running). They all have a drive for **Reproduction**, the preservation of their line of the species. They also have **Specific Attributes**, like **Movement in space and time**.

Some species are living and moving in water, some underground, some others in air. These are all specific for each individual species. Also, the **Level 1, 2, or 3** existence (Chart 11) is specific for each species — that is the specific attributes of the **Matter and Mind responses to the environment**. And finally, the **existing Sensors out of the "Six Senses"**, that also gives biological entities specific attributes.

One can see from these descriptions that "**each species is locked into their own limitations**". It is a "**very important conclusion in McKaneism**", because these facts will drive far more reaching conclusions when we analyze Humanity as Individuals and also as Societies. So the "Biological Attributes" place every species in an environment, where they can exist, and they can respond to the effects and impulses of that environment. (In other words, they can sense the energy changes in that environment.)

We also can conclude that if the directional process of that environment is "slow" in relative time, that gives the species time to adopt and change with their environment. However, if the directional changes of the energy level are not "slow", these biological entities have no chance for survival. (In other words, they will lose their "supporting environment" and they are extinct!) For example: The fish environment is water, so they can't survive on land. In the same way, human beings need oxygen for survival, so we can't exist underwater as a permanent environment (Don't think about scuba diving here because it is only an example of how a species with Level 3 responses (Chart 11) are capable of changing their existence in a different environment for a short period of time).

Part 1 - Chapter 4
Biology, Attributes of Living Entities Theory

Attributes of Living Entities Theory (cont.)

Now we are about to finish our journey in the wonderful world of biology. After we draw our chart to incorporate biology into our system, we can move on to investigate psychology and the human mind in the next chapter. As we see, so far all the previously analyzed subjects, like physics and chemistry, fit nicely into the world of biology. So far our "thought process" system, McKaneism, logically fits together. Of course it is no surprise, because this philosophy is the observation of reality.

Here is the updated chart:

Chart 12

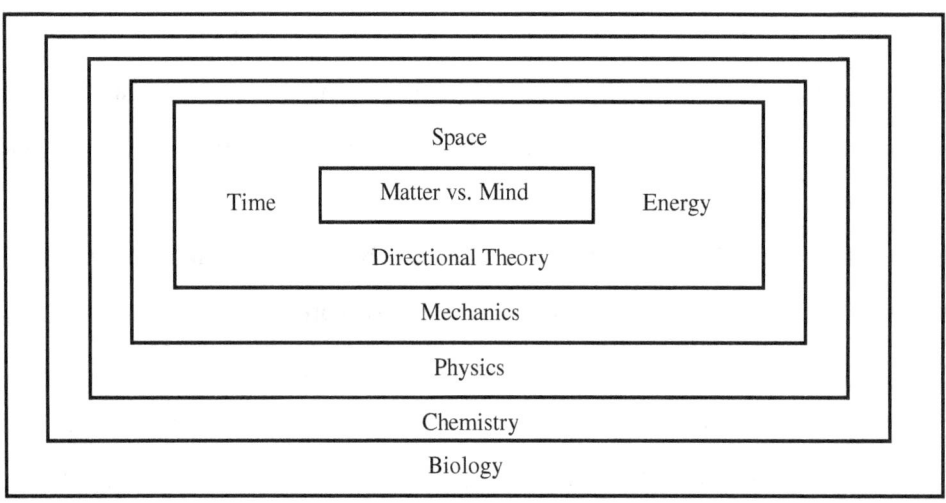

We just finished Chapter 4, so time to move on to the next subject of my Philosophy, that is Psychology. As the matter and mind complexity increases, we can see how the system of thinking and the capability of changing ones own environment will take place. So journey on, and have more fun with the following discoveries.

Part 1 - Chapter 5

Psychology, The Self-Realizing Mind, Knowledge Theory and Chart, Attributes of Human Beings Theory

Psychology, The Self-Realizing Mind

We arrived at our next chapter: **Psychology**. When I use the word psychology, I need to go into a little more explanation as to where we are really. The science of psychology studies the mind and analyzes how the mind works. It is a fact that psychology, throughout history, sometimes analyzed many different minds (like Pavlov, a Russian scientist, analyzed the behavior of dogs when they receive signals associated with food, etc.), but psychology, in the modern world, mostly analyzes the Human Mind.

We will try to discover the hidden rules of how our mind works. Also, I have to focus here on the brain surgeon doctors and psychiatrists. In our current 21st century, we have three major areas of mind study. First, the actual physicians, the doctors who try to physically repair the brain through surgery. The second are the psychiatrists, who try to alter the brain functions with chemical medications. And the third are the psychologists, who try to apply therapy through mostly verbal interaction, using the mind and it's altering power capabilities to change our health through discovering ourselves and our problems in the verbal dialogs. We also have spiritual relations with different beliefs, that suggest the functions and alterations of our mind through prayers and meditations.

As we can see, we are looking at a more complex functionality here than we addressed at the mechanical, physical, chemical or biological level. We have science and spirituality to try to claim superiority in influencing over the human mind. Just to add complexity, I will introduce a new angle, the view of philosophy of the mind. "Good old" McKaneism to muddy the water more?! (Not at all, only to give a simple explanation for what seems like a complex problem!)

The Mind is a Self-Realizing, self and environment observing existence, built on the complex formation of the matter.

What are we talking about?

The answer is simple and also very complex. In this level of complexity the matter is capable of **Self-Realizing** itself. When an animal looks into a mirror it is realizing that something similar to itself is there.

Part 1 - Chapter 5
Psychology, The Self-Realizing Mind, Knowledge Theory and Chart, Attributes of Human Beings Theory

Psychology, The Self-Realizing Mind (cont.)

The higher level of formation of mind, the human mind observes, stores the information, learns and reacts to the environment, as a definite sign of self-recognition. In short, we are dealing with an "instrument" that is able to comprehend and, in the case of humans, change the surrounding environment.

We can see it here why the many schools of thought collided throughout history — to analyze and explain the functions and existence of the power of the mind. It is almost mystical that we have a complex formation of matter capable of doing incredible things. Let us investigate, from the viewpoint of philosophy, what can be said about this phenomenon. After all, I am trying to write a philosophical publication. My observation is that the mind is formed on a very complex biological organism, called the brain. We find a brain in many of the biological existences, as we talked about in the previous chapter. However, we find that dependent on the different level of complexity of the matter is built the different level of complexity of the brain. Let us investigate this further.

Before we investigate from the viewpoint of philosophy, we need to investigate what is our current knowledge of the Mind, especially the Human Mind.

I can't be detailed here, just like other times, because the 21st century knowledge and theories around the Mind fill up libraries with thousands of books. Like always, I will try to focus on the part of the knowledge that can define the most important aspects of the study of the Mind from the viewpoint of philosophy. We find many definitions in books for the Mind, but let me go with a simple one here.

My definition is: **The Mind is a well structured, complex, living and working organism, based on chemical and biological matter (brain), that is capable to Self-Realize itself and able to observe, reflect, and respond to it's environment**.

We can see from this definition that chemistry and biology are embedded in "psychology". All the rules and knowledge of chemistry and biology apply when one studies the Mind. As we learn more and more about the Mind we are realizing that it has a hard to imagine complexity. But around that complexity we can identify simple functionality.

Part 1 - Chapter 5
Psychology, The Self-Realizing Mind, Knowledge Theory and Chart, Attributes of Human Beings Theory

Psychology, The Self-Realizing Mind (cont.)

It is interesting to see how simple functions are performed on the outside, as a result of a extremely complex work of the mechanism inside. Here I am not just talking humanity. I have no knowledge that any scientist ever analyzed the plants life to answer "does a plant have a mind?" Think about it, when we harvest a tree, how would we feel about it knowing if the tree had a Mind? It would put the tree, as we do with some animals, on the endangered species list. (It would be interesting to analyze those plants in the rainforest, that are praying on small insects!?) I can't give you an answer here from the viewpoint of science for this question but, let us investigate the functions of the mind a bit further.

Interestingly, but not accidentally, we run into the three Levels (Chart 11) that we already discovered in our chapter on biology.

It seems like every living organism has a **Level 1** (**Reflex**) response to their environment. We find this with human babies as they are ready to find the food source (their mother's breast) in the first second of their life after birth without any education or tutoring. We can observe an encoded (seems like genetically) response for that situation. Next we can see with the growth process the **Level 2** (**Emotional**) response system develops in the kids mind. After many years of continuous development (years of human schooling) finally we arrive at the **Level 3** (**Logical/Reasoning**) response system.

We are all humans that have these capabilities, even if some of us choose not to use or develop them. In this level we have the capability to observe, store information, learn about our environment, and to respond and alter that environment. In the beginning of the 21st century, the Level 3 response system is the most advanced, as far as we know about it. Very few species, other than human, show the signs of this level.

We observed only minor development, like the elephants in the wild that are capable of changing the environment, when they push over a tree to be able to reach the top of the tree for food.

Looks like we (humanity) are standing alone when it comes to comprehending the environment at the level when we are able to do critical changes in it.

Part 1 - Chapter 5
Psychology, The Self-Realizing Mind, Knowledge Theory and Chart, Attributes of Human Beings Theory

Psychology, Knowledge Theory and Chart

Let us investigate the "Knowledge Theory and Chart" next. We see in the previous pages that for humans it takes a long learning period in life before they are capable of functioning in a society. In other words, it takes a long time before they acquire the **Knowledge** needed to function properly as individuals or in the society. For that humans need the knowledge about our natural (forests, parks) and not natural (man made: society, law, etc.) environment.

What is that Knowledge?

Knowledge is the set of information about **all** subject matters. We know that no one in humanity "**knows it all**". We are learning about different subject matters, with a different deepness throughout our life. Taking in information through observation and thinking, storing that information, forgetting some, re-learning some, and on and on.

We can see, it is an ever lasting (throughout our life span) dynamic process, so when we try to create a reasonable chart of our knowledge, we need to realize that "The Chart" is only reflecting reality, with some reservations. We touched base with the viewpoint in the chapter when we analyzed energy, that the energy level of things constantly changes, so does our Mind, so does our Chart. However, this chart can represent a starting point of our analysis. Another time I will investigate and extend this Theory a little deeper.

For now, let us take an individual healthy mind from the end of high school and try to create a **"Knowledge Chart"** (Chart 13). This will be created for our analysis and representation of the subject matter, not with the intention to be 100% accurate to describe someone. Let us spend a little more time explaining the "**Knowledge Chart**" (Chart 13). In this example the square represents all the possible knowledge in high school. (For example: knowing every book, word by word, in high school in all four years and also understanding the meaning of those books.)

The bar diagrams **A** to **D** represent the subject matters (of course here I picked only four, for the representation of this example, we all know that we have more in high school.)

Part 1 - Chapter 5
Psychology, The Self-Realizing Mind, Knowledge Theory and Chart, Attributes of Human Beings Theory

Psychology, Knowledge Theory and Chart (cont.)

The height of the bar diagram represents the knowledge level of the student in each subject matter.

<u>**Chart 13**</u>   <u>**The Knowledge Chart**</u>:   Maximum possible level of knowledge.

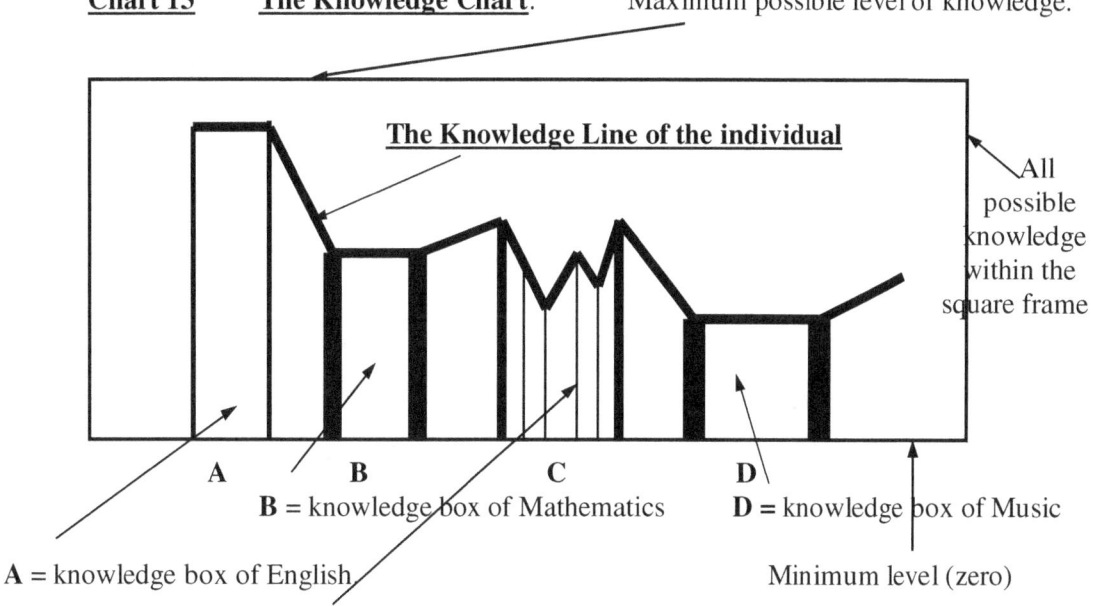

The Knowledge Line of the individual

All possible knowledge within the square frame

A   B   C   D

B = knowledge box of Mathematics   **D** = knowledge box of Music

A = knowledge box of English

Minimum level (zero)

C = knowledge box of geography, broken down by knowledge of countries

For our example we break up subject **C** to illustrate that each subject can be analyzed as detailed as we choose. (For example: in **C** the knowledge of each country within the knowledge of geography.) Finally: **The Knowledge Line** of the individual represents the achievement in high school compared to the maximum and can be compared to each other. (In this chart we see an individual with high level of achievements in English compared to the other subjects, that also can be a good indication of career choices.) One can see that based on measurement, this chart can be a very precise representation of the knowledge of the individual. Also we don't have gaps between the bars on the chart. For example between **A** and **B** we can have **A1** = physics, or between **B** and **C**, **B1** = history, etc.

Part 1 - Chapter 5
Psychology, The Self-Realizing Mind, Knowledge Theory and Chart, Attributes of Human Beings Theory

Psychology, Knowledge Theory and Chart (cont.)

With those changes our chart will look like the following:

Chart 14   The Knowledge Chart: (Extended with A1; B1; C1)

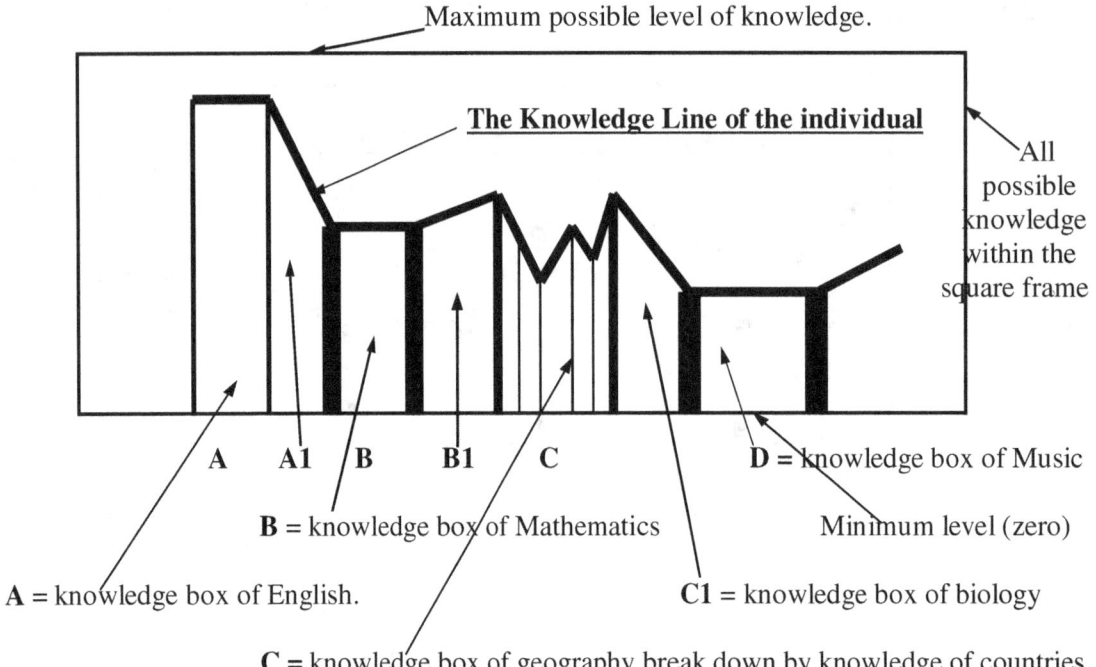

We can apply this kind of chart at the four years of College level (example: BS) that includes the knowledge of High school, then add two more years of College or University level (example: MS) including the knowledge of the four year schools, etc.

The "Knowledge Theory" is analyzing how we human beings observe, learn, store information, and think from a philosophy point of view. The "Knowledge Chart" can be a visual representation of the sum of our acquired knowledge. It can visually represent acquired knowledge that, as the human mind gains knowledge by quantity, it can change the comprehension and functioning by the quality of that mind. We can observe also that each individual, at each given time, has their own capacity that can be represented on the chart. If we look at it in a time period we can represent one person's knowledge with a series of charts in that time period.

We see the sample chart, but What about the Knowledge Theory?

Part 1 - Chapter 5
Psychology, The Self-Realizing Mind, Knowledge Theory and Chart, Attributes of Human Beings Theory

Psychology, Knowledge Theory and Chart (cont.)

The "Knowledge Theory" of McKaneism is built on the observation of humanity. It looks like we are all born with the Level 1 responses, but we learn and develop, interacting with our environment, the Level 2 and Level 3 responses (Chart 11). If it looks like I am trying to categorize humanity into some sort of "Knowledge Level" system, I am not, humanity does. Each and every individual set their own deepness in the Levels based on the effort that each put into his or her education and personal development. It is a lot more than schooling. Because the school systems didn't read this book yet (for obvious reason: I am just writing it), they are not paying attention for personal development, they just deliver a knowledge and everybody can pick it up if desired.

The personal development is far more broad than simply learning a few subjects in the average level. One of the most complete personal developments was applied in the past in Japan in the "Samurai Schools". They perfected a system that provided many aspects of personal development, from science to art and the art of their "philosophy" to the martial art; they extended the teaching all around. Of course that system won't fit the current 21st century requirements of the world, but can be viewed as a model to develop a more comprehensive personal development system.

As we (humanity) mature in life as individuals we start building our emotional system. In our emotional system we get influenced by people we are emotionally close to and admire. First line of emotional contact is our parents, then teachers in schools, perhaps priests in churches (respectively for the family religion), and then relatives and friends around us. Humanity does not realize it in the current 21st century, that in this stage, one can close a young mind, setting it up in one direction or another.

For example: if someone grew up with the strict code of Christianity, that person has the greatest possible chance to stay that way the rest of his or her life. Based on the influence of that environment and the emotional closeness, this person won't be able to be open minded for any other thought process.

The Catholic Church itself, in the 14th century, burned people that believed the Sun is the center of the Solar System (not the Earth?!).

Part 1 - Chapter 5
Psychology, The Self-Realizing Mind, Knowledge Theory and Chart, Attributes of Human Beings Theory

Psychology, Knowledge Theory and Chart (cont.)

More than 350 years later, the Pope admitted that Galileo was a good Christian, but not that he was correct and that the Sun was the center of our Solar System.

The "thought system" we believe, know, or live by serves as a screening system in front of our eyes (like a screen to look through); it can determine the individual outlook on every aspect of life. That can raise a very, very serious concern about what kind of personal development system the world lives by, and what kind of way we want to develop individuals in our societies.

The next step is a difficult one to move onto, from a highly emotional development to a Level 3, **Logical/Reasoning** response system. There is no clear dividing line between human emotions and logical thinking. Both interact and interchange constantly in every minute in our life. However, we can observe that some personal development systems concentrate on learning to control the flow of human emotions.

In the most current society in the 21st century, emotional behavior is encouraged and used by marketing and advertising. Most consumption based societies try to capture the emotions of their consumers. To have an example for better understanding, when one goes out to buy a car in the developed western societies, the emotional temptation is on the highest level. The presentation in the showroom, the sales person at the dealership, the brand recognition from the TV advertisement, the peer pressure of neighbors, family, friends, and relatives creates a very high emotional state for this consumer. However, the pragmatic view of shopping that most of us ask, "What can we afford to buy?", brings those emotions under some control and then this helps our realistic reasoning system. One can see the dual existence, at the same time, between the Level 2 and Level 3 systems (Chart 11) within ourselves.

Another example is at this moment when I am writing this book, I feel very emotional (compassionate) about how humanity neglects "Philosophy as a science", yet in the other hand, my side of emotions and my reasoning mind tell me that I will open a can of worms with this publication and maybe not for the benefit of myself.

Part 1 - Chapter 5
Psychology, The Self-Realizing Mind, Knowledge Theory and Chart, Attributes of Human Beings Theory

Psychology, Knowledge Theory and Chart (cont.)

What can I expect from those individuals of humanity who for whatever their own reason, spend more than 350 years and still don't admit the simple fact that the Sun is the center of our Solar system, even if the knowledge was available throughout the entire more than 350 years?

In this publication I am going much further than any of my predecessors ever went. To tell humanity that it is time to understand ourselves and build a planet Earth around us that makes sense, that it is a challenge to take on the whole planet Earth's population. After many, many years of thinking I feel very strongly that humanity needs this publication.

To understand and view reality, sometimes we need to control our emotions, but very few of us is willing to do this. For reasoning thinking, when I tried to introduce when I was simply stating that we don't have the proper knowledge to decide Matter vs. Mind, many of my critics will have a very strong emotional response. (Most will think I am lacking the knowledge.) To decide how sure you can be about yourself, just keep reading, because more surprising sections are on the way. Next I will address a new idea to analyze the: **"Attributes of Human Beings Theory"**. We will focus on humanity and its boundaries and limitations, incorporating the previously discovered attributes of Biological existences.

Psychology, Attributes of Human Beings Theory

The **"Attributes of Human Beings Theory"** is an extension of the **"Attributes of Living Entities Theory"**. (I call each of them a "Theory" because they are based on simple observations, and drive from those simple observations. Some thoughts seems so self-explanatory I can call them axioms, principals, or conclusions, but I am planning to get into this deeper some other time. For now we leave it as "Theory".)

No surprise, we are living entities just like plants or animals, but our Mind and the functions of the Mind make us different. These differences are the extensions from one theory to the next, of course embedded, like always, to the previous one.

Part 1 - Chapter 5
Psychology, The Self-Realizing Mind, Knowledge Theory and Chart, Attributes of Human Beings Theory

Psychology, Attributes of Human Beings Theory (cont.)

We said before in the "**Attributes of Living Entities Theory**" that each living entity has some of all of the following attributes, (just to refresh our memories):

1. Self Preservation, 2. Energy Consumption, 3. Reproduction. (Continuation of the species), 4. Capability of movement in space and time, 5. The matter and mind responses to the environment.

The senses are: 1. sensors for Seeing, 2. sensors for Hearing, 3. sensors for Smelling, 4. sensors for Tasting, 5. sensors for Touching, 6. The 6th sense?

Level 1, Reflex response to the environment, Level 2, Emotional response to the environment, Level 3, Logical/Reasoning response to the environment (Chart 11).

Why is it so important to analyze these attributes?

First of all, Human Beings have all of those Attributes that we described above, and as an Extension, a few more!

To understand ourselves and our impact on our environment we have to know our attributes. (It is a must!) These attributes are the foundations of our existence and our behaviors. In this realistic observation, we can have very important conclusions about our lives. So let us begin the investigation of our "**Basic Attributes**" and "**Extended Attributes**"!

I will call "**Basic Attributes**" those that we defined in the **Attributes of Living Entities Theory.** We will try not to revisit these investigations, as we already did in the previous chapter, however, we will refer to them as a factual, observed, and existing base. We will spend time first investigating our "**Extended Attributes**", then we will enter our world of thoughts, where we will finally see the reason for all of this is to investigate the "**Boundaries and Limitations of Human Beings**".

"Attributes of Human Beings Theory" is the result of our investigation of the "Basic Attributes" and "Extended Attributes" of Human Beings with their "Boundaries and Limitations" — all extensions of the Attributes of Living Entities Theory.

Part 1 - Chapter 5
Psychology, The Self-Realizing Mind, Knowledge Theory and Chart, Attributes of Human Beings Theory

Psychology, Attributes of Human Beings Theory (cont.)

What are those extended attributes? The extended attributes come from the capabilities of the human mind. The working human mind, (Chart 11, Level 3 Logical/Reasoning) through the sciences, slowly extends our biological attributes. Let us take a look at them, one at a time, and analyze them. Do they stay the same or get extended?

Self Preservation, is targeted with our medical sciences. In the same way that we find self-preservation in the animal world, we actually score the worst. In the civilized societies, we don't need to be as alert, for example, as a grazing animal in the African planes has to be. In the level of societies, we will investigate later that self-preservation of the society as "the population of Planet Earth" is not that much in focus. However, our medical science, with the increasing technology available to repair and exchange damaged parts of our human body, is preserving and extending our natural length of life. Medications also serve the same purpose.

Energy Consumption, which is in other words, eating and drinking, gets a huge extension, compared to the other existences. Human beings are able to cultivate and produce a great variety of food and drink, varying the taste and nutritional values greatly.

Reproduction, again, just like our self preservation, medical science is performing more and more wonders to extend the possibilities of the continuation of the human species.

Capability of movement in space and time, is one of the most extended attributes of humanity. We are able to move in our Earthly environment with cars, trains, ships, and airplanes. Reaching to space with our rocket and shuttle technology. As far as time goes, with the speed at which we can close distances, we've changed a dimension of time. We can travel around our planet, and within a reasonable amount of time, can access all continents.

The matter and mind responses to the environment, is changed because our extended senses (that we will discuss later) can respond to our environment in a more extended way. For example: we can see through a microscope all the bacteria that try to invade the human body and workout defensive medicines against them.

Part 1 - Chapter 5
Psychology, The Self-Realizing Mind, Knowledge Theory and Chart, Attributes of Human Beings Theory

Psychology, Attributes of Human Beings Theory (cont.)

The sensors of **Seeing**, simply our eyes. With the "simple" invention of eyeglasses, we can see longer throughout our lives. And with the technical invention of microscopes and telescopes, we can see the very small and into space very far.

The sensors of **Hearing**, simply our ears. The invention of radio telescopes, attached to a computer system, allows us to "hear" at much longer distances and at more differentiated frequencies.

The sensors of our **Smelling** and **Tasting**, have their own extension with the extension of the variety in our food and drink supply.

The sensors for **Touching**, as far as I know, do not need to be extended too much. Of course, as the number of objects we use in our daily life extends, our senses of touch experience more and more different shapes and sizes in those objects.

The sense that we call the **6th sense**, is still a questionable sense. Most of us don't need this sense in our modern life. The explanation is that the 6th sense (the sense of feeling the changes around us) is still questioned. Some scientists believe that we had this sense in our very early development as humans, then we lost it. Let us leave it there for now.

(Chart 11), **Level 1, Reflex response to the environment**, is the same in humans probably from the time of the first human being.

Level 2, Emotional response to the environment, is extended beyond imagination by the 21st century. A lot of our motivation as a human being comes from emotional drive. Especially in the more advanced capitalist societies, where the healthy functioning of the society is consumption based, the emotional inducement to feel good about things is incredible. We can find this from the advertisements of products, to church services, to election campaigns. We can conclude that Emotional orientations drive most of the events of the world in the beginning of the 21st century.

Part 1 - Chapter 5
Psychology, The Self-Realizing Mind, Knowledge Theory and Chart, Attributes of Human Beings Theory

Psychology, Attributes of Human Beings Theory (cont.)

Level 3, Logical/Reasoning response to the environment, is also extended greatly by the 21st century. However, as much as science is expanding in the human existence, the number of humans in the science community stays low compared to the size of the human population.

The emotionally charged responses move out the logical/reasoning responses in most human beings minds. We will spend more time to analyze this situation later.

So far, we analyzed all the attributes that we introduced in the **Attributes of Living Entities Theory,** and we see their extensions compared to any other existing living thing that we know on our planet. (Unfortunately, we don't have any other place to observe yet, just our own Planet Earth!)

Now we will move on and investigate **Attributes** that are very special and that we find only in the existence of human beings as we know it in the beginning of 21st century. These are the real and special attributes that extend and separate the **Attributes of Human Beings Theory** from the **Attributes of Living Entities Theory.** Let us investigate the last subject matter of this chapter, the extensions of human attributes. Then spend a little time with boundaries and limitations that humanity has, but for emotional reasons, many times greatly ignored.

The most different attribute of humanity is the capability to change and alter their environment. One can say that animals can do the same, for example, when army ants march through the forest and eat every other living animal that does not step a side, they are altering their environment. That is a true statement, however, no plant or animal species, only human beings, are truly able to create "artificial" = "non-natural" environments on their own. Our (humanity) alteration of the environment is not within the rules of natural interaction with the other species if we choose that way. The army ants did not have that choice. For example: no plants or animals can light a fire, as we can do, with a match. No plants or animals can build a space station in space circling Earth as we (humanity) did.

We can create a piece of land to cultivate, locking all other existence out of it, or we build homes for ourselves that are off limits for any other existence, if we decide.

Part 1 - Chapter 5
Psychology, The Self-Realizing Mind, Knowledge Theory and Chart, Attributes of Human Beings Theory

Psychology, Attributes of Human Beings Theory (cont.)

It is a very important attribute of humanity that a working, thinking mind is able to create an artificial environment around ourselves. We will make an argument later that humanity is guided by the emotions (and ignoring logic), for example, building homes for a nice view in places where heavy rain can wash them away.

But one way or the other, the statement stands that humanity can create their own "artificial" environment. This extended attribute is specific for humanity and we didn't observe any other existence yet that is capable of doing the same.

The "**Attributes of Human Beings Theory**" won't be complete without analyzing the other side of our extensions, the Limitations and Boundaries of Humanity.

We human beings require a certain type of environment to exist. Here I use **Limitations** as a **force against our** constantly extended **capabilities** and not in any negative way. The environment that can support us also contains the **Boundaries** for humanity. We need a **shelter**, because we can exist only between a certain interval of temperature, **water** to drink, and **food** to eat as an energy source. Of course, the most important is **air** to inhale.

We are **land based** and can't live underwater or can't fly in the air. (Of course, we use our inventive mind to extend our *capabilities* creating submarines and airplanes, but that won't change our basic attributes, only our capabilities.) When we create our artificial environment we satisfy all our requirements for life, keeping our mind focused on our boundaries and limitations. Also we need a certain size of land to support each individual human. (Unfortunately, I can't say we are focused on that in the beginning of the 21st century.)

Finally, the **Attributes of Human Beings Theory** is a relative theory, where humanity keeps pushing and extending our Limitations and Boundaries, using the powerful discoveries of the human mind. That is another attribute that only humanity has. No other species is able to extend beyond their natural boundaries, only human beings.

We just finished Chapter 5, so time to move on to the next subject of my Philosophy, that is Sociology.

Part 1 - Chapter 5
Psychology, The Self-Realizing Mind, Knowledge Theory and Chart, Attributes of Human Beings Theory

Psychology, Attributes of Human Beings Theory (cont.)

The matter and mind complexity will increase again, and we will investigate and analyze how this existence, "Humanity", forms social structures from the individuals; and what kind of limitations, boundaries, and extensions of capabilities society has that we didn't see until now throughout our discoveries.

Before we do that, let us incorporate Psychology into our chart:

Chart 15

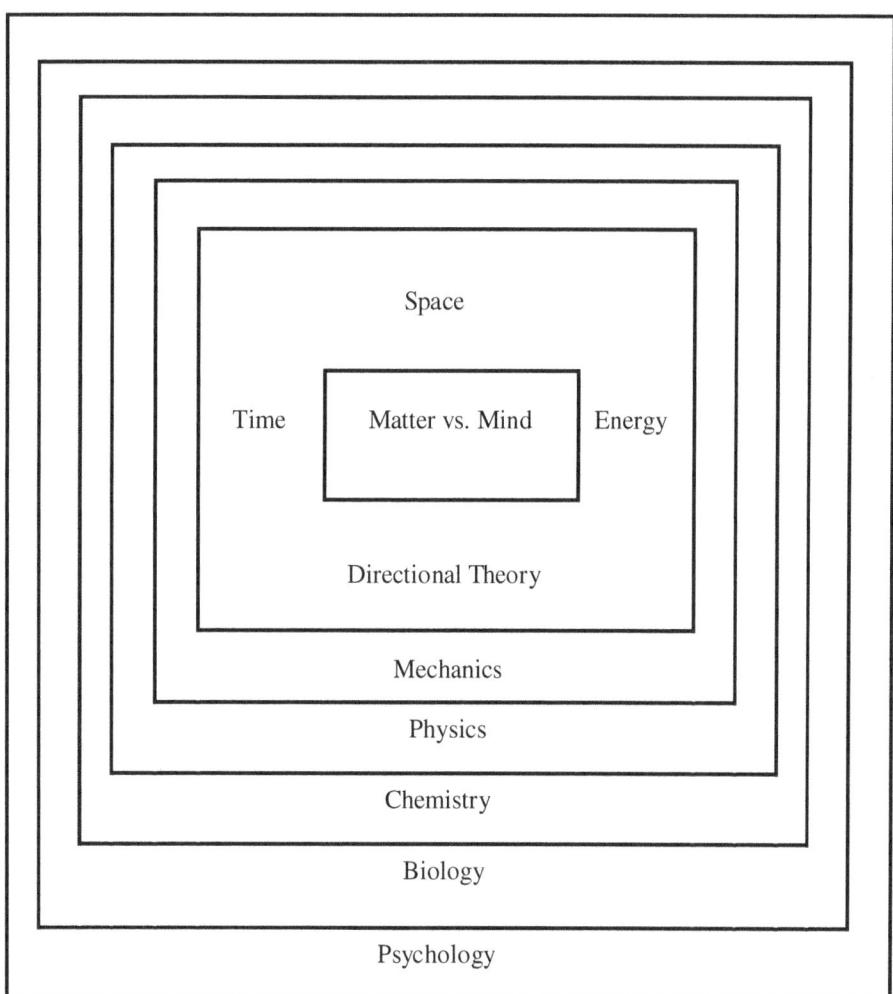

Part 1 - Chapter 6
Sociology, First Level, Fair Share Distribution System Theory

Sociology

I am using the word Sociology, as I did in the previous chapter, with the word Psychology. It is covering a broad collection of sciences and belief systems. (For example: History, Economy, Aesthetics, Law, Ethics, Christianity, Islam, Judaism, Hinduism, Buddhism, etc.) We talk about sociology when we are viewing individual human interactions. We did this generalization under Psychology in the previous section to investigate the individual Mind, specifically the human Mind. Moving to the next, more complex appearance of matter and mind, we reach the interactions of many-to-many individual matter and minds.

Before going any further, I need to stop for a few comments. The next two chapters, 6 and 7, will cover the topic I summarize under Sociology. In Chapter 6, I will investigate the basic foundation of social infrastructure, that is the "First Level": Production-Distribution System of the society. In Chapter 7, I will spend some more time investigating the "Second Level": Economy, Law, Ethics, Morale, Aesthetics, Religions, Art, Politics, Education.

The following Chart 16 offers an easy visualization:

Chart 16

Second Level; Economy, Law, Ethics, Morale, Aesthetics, Religions, Art, Politics, Education
First Level; Production - Distribution System

I will argue in these two chapters that the organization of the "First Level" of the society provides a foundation and determines the "Second Level" of existence and structure. I will look back in history and bring many historical examples and facts into this investigation.

Part 1 - Chapter 6
Sociology, First Level, Fair Share Distribution System Theory

Sociology (cont.)

I will try to highlight some of those facts that I think make it easy to observe, as well as realistic and factual, and to validate my reasoning and conclusions. Like I've always said, if you don't believe me, read and check the facts, and then believe in the discoveries of your own mind. I can only show you the road, it is up to you to travel on it.

Now we are ready to continue our journey into the next two chapters to analyze the next level of complexity of matter and mind; their actions and reactions in the environment of sociology.

We observe social behavior in forms other than human existence. Many animals are living in social groups. (Like monkeys, elephants, lions, etc.) Just like before, my viewpoint is that these groups totally depend on nature. For example, if we have a great drought in Africa, the whole herd of elephants can die out. They are not able to change their environment, only adjust to it, and they can do this only up to a point. Only slow, many hundreds to millions of years of, evolution can make any species able to adjust to environmental changes, with the exception of humanity. When we (humanity) have a drought, we can store water in good years, or alter the path of the rivers to irrigate the land, if we want to do it.

In "The Foundation of Humanity" we will focus on basic sociological rules that define the interaction of individuals. We will discover here why it is so important to understand the individual, because some of the individual drives don't change just because they form a society. When we understand individual drives, needs, and behaviors we can incorporate those into social interactions and behaviors in societies.

The social foundation of humanity, in the philosophy of McKaneism, is the First Level system of **Production** and **Distribution**. A long time ago, in the history of humanity, the production of things to serve our needs was broken up to a group of individuals. What am I saying here? It is easier to understand through some examples. In the human societies, we have people (or as we called individuals) who grow crops for bread, or make clothing, or build bridges, or assemble cars, etc. I said before that the jobs to support the needs is what is separated between the groups of individuals. That is the production side of the society.

Part 1 - Chapter 6
Sociology, First Level, Fair Share Distribution System Theory

Sociology (cont.)

The production takes place because we need to support our basic individual and social needs to stay alive and have a healthy, functioning life. We investigated in the previous chapter that we need shelter, food, drinking water, and a few more things to function and survive in our society. Very important to note here that regardless of the social structure, (slavery, feudalism, capitalism, socialism, etc.) we need to have those items produced that support the survival of the individuals.

Here we are in the society and we already encountered one of the very important cornerstones of our existence in the society, that is **Production** of our goods and services. It is not an easy step for the society. Producing some of the needed items can be a very hard, very trying experience for those individuals who decide to do it. For example: working on the fields fourteen to sixteen hours a day to produce crops for food, this is definitely hard work. Production is more complex in the most advanced societies, and less complex in the more simple ones. In our 21st century we've encountered and observed tribal structures in the Amazon rainforest that are still living in the first man-made structure of society, which is the "Tribal" "Social Structure".

Their production partially takes place by nature and that is providing some of the naturally harvested food items from the forest. The man mostly hunts, the woman mostly gathers the food. Handcrafting of personal items such as jewelry, and some tool making, become a real production activity in their environment.

I noticed that some people "talking big" saying "There is so much out there!". It is a very amusing chat (maybe) at a social gathering, and may seem to make the talking person very mysterious and knowledgeable, but it is an empty talk. Simply, in the human societies we have production functions based on needs (So much out there). If we have a human need for some product, it gets invented by individuals and introduced in the market for satisfying the human needs through consumption. That happens again and again, regardless of the social structure. This discovery can lead us to the next important rule of societies, that is **Distribution**. All these products that support our needs need to be distributed, because that is the main driving force for their production.

Part 1 - Chapter 6
Sociology, First Level, Fair Share Distribution System Theory

Sociology (cont.)

Without being used by individuals in the society, their production can't be justified. The use of those products (and of course the purchase in the market place in modern economies) justifies their existence. Products that don't become part of the distribution system disappear from the market place and soon the production of them will disappear as well. Here we are, the society revealed. It is simple like that. The unchallengeable truth and reality. Look around yourself, observe, then see if you do not draw a similar conclusion.

So far so good, we can say that. We have **Production** and **Distribution**. Can we challenge the statement that these two look like "simple" functions in the society that determine all the other interactions within all the individuals in the society?

Of course, anyone can challenge this statement, but I have a news for you, you will lose the challenge. Simply put, individuals can't exist without satisfying their basic needs, so they **have to produce (Production!)**. More than that, to get those products to each individual in the society, the system of **distribution has to exist (Distribution!)**.

I can say, case closed, the challenge is over. If anyone feels or thinks that this reasoning does not make sense, that "one" needs to spend countless hours in reading the previous chapters and the first few pages of this chapter, and then just sit back and sort out in one's own mind the thoughts until finally "one" will see the light.

I'll throw in a few examples that we will revisit later in this publication and that we must analyze deeper. For example the Law of the ("land") society or the Ethics of the society are also the matter of distribution. Think about it for a minute: who creates the Laws, and who sets the Moral codes for the society.? In the USA (in the beginning of the 21st century) people who make a law in the House and Senate are people who get elected because they have millions of dollars of resources to run in the election process. (I have to say, again and again, that this is one of the best systems on Earth but not without room for improvement!) The per capita income is around $21,000 per year, and these officials are voting for their own pay increase to make somewhere around $150,000 per year. Looks like they are in total control over the distribution of their wealth in this society.

Part 1 - Chapter 6
Sociology, First Level, Fair Share Distribution System Theory

First Level, Fair Share Distribution System Theory

Somehow most of the people are so misinformed or apathetic that they have no clue or don't care about these situations at all. We traveled this far, so let us continue the journey.

Anyone can observe the process (in the USA) that every morning hundreds of thousands of cars roll on the highway, taking people to work. That is simply the part, **Production**, producing all the goods and services. The same way anyone can observe that most of us go out shopping to pick up goods (food) and services (dress from a dry cleaner) that we need for maintaining our existence. That is the other part, **Distribution**, we distribute all the products and services between all of us, using money that we (as humanity) print as a tool of the exchange. (We won't jump in deep to investigate that some people don't produce but take their share of the distribution an so on, for example, like people on welfare in the USA, etc.) In the general sense, I hope we all observed and got the picture. This interaction in society (Production / Distribution) raises a few very important questions:

Is Production / Distribution a natural or an artificial system?

Who decides who should produce what?

Who decides who gets how much out of the distribution system?

If you think the answer is easy, try to answer it yourself. It takes a little more investigation than you think. If you ask "why is it important?" then the answer is this: we run into an observation, that all of us can observe every day, and it seems like the foundation of our social structure and functioning, and hence may deserve a little discovery!

We always can do one thing, so let us look at our biggest natural laboratory, "mother nature". We can see, for example, the zebra and lion contest in the natural setting in the safaris of Africa. We also know that they were around in the time of Egypt many thousands of years ago (or earlier).

They both have a limited time of existence (life span) so they had to have many generations in those thousands of years. Assuming that humans stay out of their life, their conflict is a natural process. Their existence is a "natural system". You will be surprised that it is a **"Natural", "Fair Share Distribution System"**.

Part 1 - Chapter 6
Sociology, First Level, Fair Share Distribution System Theory

First Level, Fair Share Distribution System Theory (cont.)

They both survive until they have a balance in their existence. If the lions kill more than their Fair Share of zebra and the zebra becomes extinct, then the lions food becomes extinct and so they become extinct with it. (You can say there are other grazing animals for a food source, then for example, I can say they eat them all!)

It seems like lions kill only as much as they need to survive and the natural process of reproduction keep the zebra number at the level of survival in their given territory.

One can look around nature and will find that the "natural system" of existence is well balanced. Actually in nature one can find a "food chain", that serves a purpose of balance. All the different species following the rules and laws of Mechanics, Physics, Chemistry and Biology (if you remember our chart) to have a balanced existence. If one takes humanity out of this process, one can find that this natural system can and will exist on its own even without humanity. We've found a **Fair Share Distribution System** in our natural environment where the natural reproductive life cycle does the **Production** and each species takes their consumption (**Fair Share Distribution**), never exceeding their needs.

Now we can investigate the human societies, and we can try to answer our questions from the previous page. Is the human Production / Distribution system a "natural" or an "artificial" system? The answer is: **Artificial System**.

We did see before that humanity is the only known existing species in this planet who is able to alter, change, create, and destroy their environment, sometimes with artificial forces (humanity has made, for example, an A-bomb). If it is an **artificial system**, is it a **Fair Share Distribution System**? Unfortunately, the answer is **NO!** (of course the answer is NO, if it would not be, then I wouldn't have to write this publication!). In the following section we will investigate the type of human made systems and how are they relate to each other.

Way back in history, around ten, twelve thousand years ago, humanity started to take charge in the planet Earth environment. (That is the time when all the problems started!) Not observing and understanding the rules of natural laws and processes, humanity started to use the environment to serve its needs without consideration of anything.

Part 1 - Chapter 6

Sociology, First Level, Fair Share Distribution System Theory

First Level, Fair Share Distribution System Theory (cont.)

The production started to clear forest area, using the land for agriculture and raising animals, using the trees from the forest for fire, building homes, and on and on. So the **Production** began! Now is the time to answer one of our previous questions: Who decided who should produce what? The answer is relatively simple. Production is based on the needs of the society, so that is what determines the production of **What**?

The next part of the question is who produces what; that is coming from "tradition" in some societies and coming from "free choices" in some others (like USA). More or less we have fairness in productions systems. People are using their skill and education to choose which part of the production system they want to participate in.

At the same time, societies created rules to govern the relations between the individuals. These rules evolved to religions and constitutions and laws in time.

In these rules the basic forms of **Distribution** were spelled out. In other words, who gets what and who deserves what in the society.

Who decided that who gets how much out of the distribution system?

Looking at history closer, one can realize that distribution is a very broad term. It isn't a question about distributing what we produce, because one of the basic foundations, "production", is driven by the needs of individuals (that calls for distribution). Having said that, we would think that it is easy to answer: Who decided? It is not. The distribution in the "natural systems" plays out in a very brutal way. The strong survive, the weak are out of existence, and if the environment changes to be unbearable, then everybody becomes "history", strong or weak. We have our great example with the dinosaurs where (possible?!) an instant climate change managed to kill them all. We find bones, but we don't run into any of them alive. It can be a very good thing, as much as we can understand, because of their mad behavior based on their sizes.

We must observe this because the human distribution system is human made. And it is raising a lot of questions about it. Especially, how is this "**artificial system**" defined, human made, and function? Let us investigate this "artificial system" in a little more detail.

Part 1 - Chapter 6
Sociology, First Level, Fair Share Distribution System Theory

First Level, Fair Share Distribution System Theory (cont.)

In the history of human kind we find different forms of organization, how people lived together and shared their production. Based on the **Production / Distribution** system, we know six existing models throughout our human existence up to today.

These are the following:

1. "Natural Distribution System" ("Mother Nature") of Production / Distribution system (this is the only one where there is no human influence and it exists from the beginning of time of living entities).

2. Primitive communal (Tribal) system of Production / Distribution system (this comes from the time of the beginning of human kind, about 100,000 years ago, and "interestingly?!" can be observed 100,000 years later in the tribes of the Amazon rain forest).

3. Slave State societies system (or slavery) of Production / Distribution system (like the Greek and Roman Empires, or before the civil war in the USA).

4. Feudal State societies system of Production / Distribution system (like the 11th and 13th centuries in Europe).

5. Capitalist societies system of Production / Distribution system (like our current modern 21st century USA and Europe, Japan, etc.).

6. Socialist, Communist system of Production / Distribution system (like before 1990 Soviet Union, or 21st century Cuba, China, North Korea.)

One can argue that there are many other different systems that exist on our planet Earth, but with closer investigation, it is easy to see that they are one of these five (#2 through #6). Or perhaps somewhere between one of those five and in the process of evolving from one to the other, or maybe a mix of a few. We will investigate each of those in Part 2 of this book very closely.

For now, I will introduce a new one which is the seventh:

7. **McKaneist, Fair Share Distribution System** of Production / Distribution system

Part 1 - Chapter 6
Sociology, First Level, Fair Share Distribution System Theory

First Level, Fair Share Distribution System Theory (cont.)

Here I don't want to create a "big name" for myself calling it McKaneist, but I don't know any other fitting word for it. Anyway the name is not important. The important part is to discover a system that can be "**Fair**" to everybody who is living in it.

We can see here that each of these systems has many common features. One is that they all have **Production / Distribution**. The other is that each system has **Leaders** in the system who get control over the regulation of Production / Distribution.

I argued earlier that there isn't much need to regulate **Production**, because that will be regulated by the fulfillment of human needs for finding an occupation that the individual can and likes to do, however, it is very important to see that **Distribution** is always very regulated in each social form, by the leading group of individuals.

We don't want to call it "leading class" intentionally, because I don't want to introduce a "class" driven society.

Another very important feature of these previous systems is that **each of them gets replaced by another as time goes by**, so they are evolving (or remember earlier "Directional Theory", they all moved to a certain Direction!). In the following pages we will spend some time investigating these three common features a little closer.

First of all, each system had emerging **Leaders**, either by overpowering the rest, like in the early **Primitive communal system** or by selection of others in that society (elections) like in our modern USA. The rest of the members of each of those societies then follow their leaders or value very highly the **Direction** that their leaders point to. Those leaders became, somehow, more than an average member in every setting of those social structures, so they demand and get more out of the **Distribution** systems. Somehow the rest of the society thinks that it is OK for those leaders to get a **bigger share** of the goods and services that get produced in those societies. It seems "**Fair**" for the others somehow! (Without thinking, how much more is that **bigger share** "**Fair**"?)

Now we introduced a very important word for humanity, it is: "**Fair**". We can see easily that **fair** is a very relative term, very closely related to a human emotional status.

Part 1 - Chapter 6
Sociology, First Level, Fair Share Distribution System Theory

First Level, Fair Share Distribution System Theory (cont.)

Throughout history many changes in societies were introduced, because the society was "feeling" bad about the existing system, or in other words, the society found the existing system "**unfair**"! These corrections were system changes through **reforms** (like Amendments to the Constitution in the USA) or **revolutions** (like one that abolished the kingdom in France in 1789, or the other that changed the society in Russia in 1917 to Communism). Even in our daily life, when we find people talking, they try to share something like "50-50, it sounds **fair** to me!" — implying then that sharing something by half and half (or 50%-50%) sounds like a **fair** deal. This observation is introducing to us a very important thing: It looks like our human made distribution system was based on the human emotional status, or we can say the "Emotional status of all humans", in the given societies. It is very important to realize, for many reasons.

Let us investigate the previous and our new "recommended" (Fair Share) social structures based on our new discoveries of "emotional vs. logical/reasoning" distribution systems. (Interestingly we run into our basics from the previous chapter, Psychology, where we did see the Level 2 and Level 3 systems in humanity being "emotional vs. logical/reasoning"!) Because it is our human attributes (as we saw before) to have a Level 2 and Level 3 system (Chart 11), it is not a surprise that those levels play into our social structure very closely.

One more important discovery is that each of those social formations (remember #2 to #6) were changed in time throughout human history. We also need to note that they changed by reforms in a peaceful way or they changed through forces like revolutions and wars.

For example: Hungary changed from Capitalism to Socialism/Communism after the Word War II, when the liberation from fascism by the soviet army brought soviet communism with it in 1945. Hungary changed back to Capitalism around 1990 with reforms without any revolution or war.

Let us do some structural analysis of all of these systems and investigate the merit of all of these from the standpoint of human emotional and logical interactions.

Part 1 - Chapter 6
Sociology, First Level, Fair Share Distribution System Theory

First Level, Fair Share Distribution System Theory (cont.)

Here are the possible combinations of our investigated Production / Distribution system structures based on **Fairness** and **Equality**. **Equality** is a **Logical/Reasoning** (Level 3) evaluation in the distribution system. **Fairness** is the **Emotional** (Level 2) and/or the **Logical/Reasoning** (Level 3) evaluation part.

My argument is for us to try to make the distribution system an **Artificial Level 3**, because it hasn't happened before (yet) in human societies, however, it has existed a long time, way before humanity, in nature! Now let us view the combination of the categories of these systems, and place all the previously existed, and our new recommended system, under these categories.

Here are the categories of possible Production/Distribution systems:

<u>Chart 17</u>

Unequal Unfair	Unequal Fair	Equal Unfair	Equal Fair
2. Primitive communal system	1. "Natural Distribution system"		6. Socialist, Communist system
3. Slave State societies system	**7. McKaneist, Fair Share Distribution System**		
4. Feudal State societies system			
5. Capitalist societies system			

Let us investigate these categories (Chart 17) and take a closer look; what we can tell about each of them.

Part 1 - Chapter 6
Sociology, First Level, Fair Share Distribution System Theory

First Level, Fair Share Distribution System Theory (cont.)

The 1st system, the "Natural Distribution system", is natural and not human made and so we won't spend much time investigating it now. We will focus more on our categories, the human made systems. We can see that under the "**Equal Unfair**" category we don't have any known existing structure, natural or human made, possibly because if a system has **Equal Distribution** it is very hard to think about it being **Unfair**.

If we have five people and five apples, and each person gets one apple, that is equal and fair, so no one can say anything else about it.

The category "**Equal Fair**" includes the human made social structure called **Socialist/Communist system**. In theory, in Communism each person of the society contributes in the **Production** by their best capabilities, or better than that "Equally", and then takes **Equal** and **Fair** share of the rewards in the **Distribution system**. One can easily see why a system like this **can't work!**

Humanity exists as many different unequal individuals, by capabilities and needs, so enforcing an artificial "**Equal Fair**" distribution system in that kind of society goes against the unequal needs. (For example, this human made rule: everybody in the society gets 10 pounds of coffee and 50 boxes of cigarettes per year in an idle Communist society, even if one doesn't drink coffee, only tea, or perhaps even if someone does not smoke at all.) This was an extreme example for the representation of an idle "**Equal Fair**" distribution system, however, a very realistic representation of the perfect theory.

Two more comments:

First comment, the so called Communist societies never reached this idle level for two reasons: **One** reason was that to create a society like Communism, one needed a "super rich" society where every product has (in quantity) many times more than consumption needs, and those so called Communist countries were never be able to reach that level of richness. **Two**, the Socialist/Communist societies, like any other that has leaders, were using the theory of Communism (and distorting it for their own purpose) and living like kings in those societies, and they didn't want to apply the "**Equal Fair**" distribution ever (as the theory would require!).

Part 1 - Chapter 6
Sociology, First Level, Fair Share Distribution System Theory

First Level, Fair Share Distribution System Theory (cont.)

<u>**Second**</u> comment, for those in Capitalism who think the cold war ended Communism, I have some interesting news: Communism ended because it isn't a viable human made solution for a social structure, and it was only a matter of time in the remaining Communist countries before the people of those countries got smart and realized it, that the system they live under, basically, isn't good for them. However, Communism, as a distribution system was a **Level 3** system, in other words, a **Logical/Reasoning** solution for the distribution problem.

The category "**Unequal Unfair**" includes most of the existing social structures of humanity throughout history, including the current most widely established, Capitalism. These are a **Logical/Reasoning** (Level 3) based (**Unequal**) and **Emotional** (Level 2) based (**Unfair**) social structures of **Production/Distribution systems**. **Unequal** is proper, based on the unequal needs of unequal attributes of humans who exist in the society. However, **Unfair** is the problem that makes or will make all of these societies not sustainable.

Human nature (emotions, Level 2) demands fairness in the social structure, and the unfair societies all go away by time. (For example: one doesn't find (I hope not!) one person in the beginning of 21st century who believes that a "Slave state system" is a fair system for humanity!)

Capitalism is the most widely existing system on Earth in the beginning of the 21st century, and most people believe that the "Unfair" distribution is "A-OK!". This system is using the emotional, and also logical, argument that one who works harder deserves a higher compensation than one who works less or indeed, one who works "smarter" deserves higher compensation than one who cannot.

It is excellent, in theory, but in practice, throughout generations in the Capitalist societies, the accumulation of wealth occurs in the hands of a few individuals or families, and those who have the wealth gain control over the government of the society, so the wealth and power merges into one. Being in power, they then use that power to accumulate more wealth, and that goes on and on for many generations.

Part 1 - Chapter 6
Sociology, First Level, Fair Share Distribution System Theory

First Level, Fair Share Distribution System Theory (cont.)

The saying that anyone has a chance to be wealthy in capitalism is simply that — a chance (Yes, "anyone" can hit the weekly lotto too, but, odds are heavily in favor of you never winning the lotto).

We see from our previous analysis that the capabilities of humans are unequal. Physical and mental strength of the individuals varies widely in every society. Education, which must be purchased in capitalism, (and for a very high price in the top universities) is not available for all. We will spend a lot more time with these items in Part 2, but for now, let us just say the following: The "Natural Distribution System" as we know existed for millions and millions of years and it is still existing today (I will be "brave" and predict that it will exist as long as the biological existence will exist on Earth).

However, the "**Unequal Unfair**" social structures of **Production/Distribution Systems** are all going out of business (or did), other than capitalism. (see more in Part 2. interval rules, they exist in time intervals only). Finally, let us look at the "**Unequal Fair**" social structure of the **Production/Distribution System**. (**McKaneist, Fair Share Distribution System**) This system tries to incorporate our observations from the "Natural Distribution System".

"**Unequal**" as we investigated before — it is working because it is based on the unequal qualities of humanity. I think (and I believe) if we can correct the "Unfair" part of capitalism and change it to "Fair", we can establish a new system for humanity that comes closer to the "Natural Distribution System" in the biological world. In this way we can see that as long as humanity exists, the "**Unequal Fair**" social structures of **Production/Distribution System** will serve humanity. This system will be an "**Unequal**" or a **Logical/Reasoning** (Level 3) based plus a "**Fair**" or **Emotional** (Level 2) and **Logical/Reasoning** (Level 3) based (Chart 11) system.

If we look at this closer, the difference between the "Natural Distribution System" and the human made distribution systems is something that we can discover in the kingdom of biological world. That is that no entity goes beyond the satisfaction of their needs.

Part 1 - Chapter 6
Sociology, First Level, Fair Share Distribution System Theory

First Level, Fair Share Distribution System Theory (cont.)

We can't say the same for humanity; because of our never ending appetite for emotional fulfillment, we want more and more satisfaction in our life. So, if we can find a reasonable level of control over our emotional outbursts of consumption (including emotional ones) we may just find the solution.

Why does the "unfair" part of capitalism not work? Capitalism assumes an *"unlimited environment"* to accelerate in life for everybody. Freedom for all in every way of life. However, a closer investigation tells us that our Earth is a *"closed (limited) environment"* system, with lots of limitations. We investigated before, in the attributes of humanity, that humans need air to exist, and as we know it in the beginning of the 21st century, our planet Earth is the only place for our existence.

That is our human limitation. **We are living in a "closed" finite system, not an infinite one**. This system can support, comfortably, only a given number of humans. Today's scientists estimate this number to be around 2.5 billion, and we already have 6 billion people on Earth. That creates a lot of demand on our environment that can't be satisfied. Can you imagine if China and India, with the population of about 2.2 billion people, would reach the wealth level of the USA, where every household (about 120 million) has 2 cars per household? (Out of 300 million people in the USA, 40% is 120 million households, they would own 240 million cars!)

That can lead us to a calculation that with 2.2 billion people, 40% would then be 880 million households, with 2 cars per household, and you would have 1.76 billion cars in China and India. Right now (2003) the USA has some energy problems already looking into the future while thinking about finding the oil we need.

Can you imagine 1.76 billion *more* cars in this planet? (How about 1.76 billion trucks and SUV's?!) Of course I left out the developing world countries, who also want their share of the pie, like Central and South America, Africa, and some of the other nations of Asia. One can see if we already have problems on Earth with our current use of the planet, what these types of additions can do to it! I am not even entertaining the fact that within the next 25 to 50 years, the population on this planet could jump to 10 billion.

Part 1 - Chapter 6
Sociology, First Level, Fair Share Distribution System Theory

First Level, Fair Share Distribution System Theory (cont.)

It is easy to see why we need to change to an "**Unequal Fair**" social structure of a **Production/Distribution System**. **Specifically, a McKaneist Fair Share Distribution System**, to ensure the possible survival of humanity. Closer investigation can reveal that we are already in a "too late" mode with many of these changes. **The reason for that is because humanity is governed by the "feel good emotions" in their decision making, and those who try to use Logical/Reasoning or thinking to solve the problems don't have a voice.** That has to change real quick if we want to see positive improvements in the life of humanity.

It is possible, and actually most likely, that our own survival depends on our decisions that we make now and in the future. I will spend a great deal of time, or more than that, a whole chapter of recommendations for changes to make the future better in Chapter 9, and then a very detailed investigation of the "Fair Share Distribution System" in Part 2 of this book. However, these recommendations will only be to spark a thought process of humanity and try to bring together the people who can solve the problem. It would be very, very foolish of me to think that I have a solution for all the people on the planet Earth.

McKaneism is intended to be a starting document to discovery and establishing "The Foundation of Humanity" — but humanity has to jump in, pick up these thoughts, add to it, alter it, criticize it, and revise it to end up with a well rounded final version that we all really can use as "The Foundation of Humanity" + "The rest of the Human Castle".

There is no question in my mind that we need one now! We must look at the greater picture and start working on it, to live by a system that can support humanity and all our decedents many centuries to come. We can see that capitalism won't do the job. If you still have any question about it, just look around and truly think about it. (For example: knowing the law of equilibrium in a macro economy, anyone can easily see that as of today, a country like USA, that has 5% of the population of the planet (300 million) is consuming 75% to 80% of all the goods and services. That can't be sustained in the planet, and can't be expected to be approved by the other 95% of the population (5.7 billion) of Earth.).

Part 1 - Chapter 6
Sociology, First Level, Fair Share Distribution System Theory

First Level, Fair Share Distribution System Theory (cont.)

These unsustainable inequalities that were approved and created by capitalism can't lead planet Earth's population into a prosperous future for all. It is amazing how all these things fall into the right place within my system. If we apply the Directional Theory here (saying that the society, as we want it or not, is going in a direction, just like feudalism changed to capitalism), and we apply the change in quantity (growing population on the planet), that leads us to a change in quality, then the conclusion is simple: Capitalism has to go, it has outlived it's usefulness and it will change by us peacefully with reforms (my hope that USA leading the planet into a more prosperous existence as a Superpower as of year 2003), or we can wait until the "social structure" changes by revolution or changes induced by war.

Many systems changed after WWII, and we risk that our Superpower may go down with this change, as it happened with the mighty Roman Empire in the past (if you do not truly remember history, you are doomed to repeat it). This change is not optional, it is coming like a freight train, and we are standing on the railroad crossing.

Make no mistake about it, I don't want to create fear in anybody, I'm just drawing logical conclusions on the presented theories and thought processes. If someone is still skeptical, that is OK, then we will move on to Chapter 7 — where I will investigate the "Second Level" of social structures (Chart 16) in our capitalistic society to discover more clearly that more signs point to my previous conclusion, that capitalism is standing on shaky ground and it does not support us and our children's future.

Anyone can question the structure of "First Level" and "Second Level" Theory. I will return to this subject in great detail in Part 2 in this book. I will investigate and think over how the Production-Distribution system is really a base for the "Second Level", which is used to determine the additional relations between the individuals of the society. Right now, however, I have to go on with not too much explanation and more of just the observations to show the structure of an Unequal Unfair social structure — for this discussion, I observed Capitalism, and it has many problems in trying to support and care for the population on Earth.

Part 1 - Chapter 6
Sociology, First Level, Fair Share Distribution System Theory

First Level, Fair Share Distribution System Theory (cont.)

The time is now (even if we are a little late) to think and act on a new design that can take humanity into the next millennium. Before we go on to the next chapter, I have to remind all of my readers that most of my examples will come from the structure of the USA.

This isn't intended to criticize the USA, as many times I must repeat, that when it comes to living standards the USA, it is still one of the leading countries. We will see later that, unfortunately, we have countries and nations in this planet where they simply are not worth living in. I am not naming any countries now, but I will be more direct in Part 2 of this book.

The reasons I am investigating the USA are because first of all, I am living here, so I see every day events, and second because this is one of the most advanced states of Capitalism on the planet Earth, thereby making it a good place to observe and discover the rules of Capitalism in daily, practical life. Let us move on to Chapter 7 and check the rest of the problems that relate to the Second Level but build on the foundation of the First Level. (Chart 16).

Part 1 - Chapter 7
Sociology, Second Level, Economy, Law, Ethics, Morale, Aesthetics,Religions, Art, Politics, Education

Sociology, Second Level, (Economy, Law, Ethics, Morale, Aesthetics, Religions, Art, Politics, Education)

In this chapter we will see how the First Level (Production-Distribution System) determines the Second Level (Chart 16). Just like we talked about the foundation of the house, we have to go back to the same example. Capitalism is an Unequal Unfair individual based society, as a Production-Distribution system. It may sound critical, but it isn't. Anyone can look around in daily life and see the facts. In a society where you find billionaires, who have it all, and also others, who are sleeping under a bridge and have nothing, defines the structure of the society. For the people who grew up in these societies, this doesn't raise a question most of the time. They hear that most of the time that the people who live under the bridge don't want to do better, it is their freedom of choice, and in the free society everybody has a chance to achieve anything, even the billions. Sounds fantastic, so we can close all the arguments and go home happy!

Are we ?

Not so fast! All this "big idea" thought process is coming from the fact that the wealth of the planet is so big, it is hard to comprehend. Just like the speed of light, if you don't travel it you don't experience it. However, no matter how big the wealth of our planet, the number of people trying to share it is just as big. We will spend a great amount of time in Part 2 with this issue in this book. Here, just for a preview, the catch is that the planet and it's resources are a closed system — it isn't unlimited. So, when one takes a lot then one leaves little for others. The big talk and the constant reinforcement that you will be "**The One**" keeps society focused on that we all can have it (wealth) if we work hard for it. It isn't correct.

I constantly return to the Fair Share Distribution System of McKaneism, because you will see (in Part 2) how the natural law of equilibrium is moving the societies closer and closer to a Fair Share Distribution System.

The Economy, Law, Ethics, Morale, Aesthetics, Religions, Art, Politics, Education are just as Unfairly and Unequally distributed in Capitalism as the First Level Production-Distribution System.

Part 1 - Chapter 7
Sociology, Second Level, Economy, Law, Ethics, Morale, Aesthetics, Religions, Art, Politics, Education

Sociology, Second Level, (Economy, Law, Ethics, Morale, Aesthetics, Religions, Art, Politics, Education) (cont.)

No surprise because the Production-Distribution System is the foundation of the building for the Economy, Law, Ethics, Morale, Aesthetics, Religions, Art, Politics, and Education (Chart 16) — they are the rest of the building — first floor, second floor, (bedrooms, living room, kitchen, etc.), if we use our single family home example. Many times I used the magical escape, which we will revisit in Part 2 of this book, but for now you better believe me for this is true. In Part 1 I've tried to be easy to understand to my readers so I chose to use some examples instead of logical arguments, historical references, and other philosophers that many of you may not even heard about yet. Philosophy isn't a dinner table conversation topic, like football or baseball or basketball, so I've tried to address those readers in Part 2 who maybe take "Philosophy" to a dinner table conversation from time to time. Now let us spend some time with some examples for better understanding of what is my reasoning to say that Economy, Law, Ethics, Morale, Aesthetics, Religions, Art, Politics, and Education are just as Unequally and Unfairly distributed, just like everything else in Capitalism.

We all know that the name Capitalism comes from the word "Capital", which is short for the "working money".

When one has extra "working money", "Capital", one tries to find a way to make that money work and return more money then one started with. That is Capitalism in a nutshell. So far nothing wrong with that. The problem begins, when individuals in the society using the unequal unfair distribution to get too much of the capital. The world is changing for them at this moment. Here I don't debate this (you already know, Part 2!), just provide some examples.

Economy, is the constant cycle in society moving from stronger to weaker then stronger from weaker, etc. One who is wealthy is absolutely unaffected by these cycles, however, on the more modest level of living (middle or lower class in the USA) people can lose their life's worth of work and wealth (and not by choice) as part of the economic cycles. This doesn't sound fair to me, but I'll let you think about it.

Part 1 - Chapter 7
Sociology, Second Level, Economy, Law, Ethics, Morale, Aesthetics, Religions, Art, Politics, Education

Sociology, Second Level, (Economy, Law, Ethics, Morale, Aesthetics, Religions, Art, Politics, Education) (cont.)

Law, and some of those who seem above it. Doesn't it puzzle you that some privileged wealthy people can manipulate the law, using a powerful attorney and lobbying services, and that an average citizen can't afford the same financially?

What about Congress in the USA that can vote yes for their own pay raise? It is human nature not to say "no" for more money, but shouldn't it be a law that governs the process for raising their income when they deserve it, like the raises we all get, rather than the privileged ones voting for their own well being? It seems like someone is above the law. Can you vote for your raise at your work place — and actually get it ?

Ethics, Morale, should be defining the way of our standards, however, in capitalism the moral standards are unfairly overwritten by money, the main engine of the capitalist societies. We hear stories about how some of the rich ones break the moral codes and get away with it.

Aesthetics should be defined as the science of **Art**, where one can view art through the way of science. It is easy to see that in a money driven society the influence is money and not science on Aesthetics, and in the same way, money and not quality influences Art.

Religion is highly divided in capitalism, with many different beliefs. They can reach the level of exclusiveness when one believes one system excludes the others.

Politics is the manipulation of human emotions through words. The wealthy people's playground. They buy the writers, the specially trained ones, to construct the powerful politically correct (Not Philosophically correct) speeches and views to manipulate the masses.

Education is simply more money, which leads to a better education, better schools, and better possibilities in one's life to succeed in the society.

Part 1 - Chapter 7
Sociology, Second Level, Economy, Law, Ethics, Morale, Aesthetics, Religions, Art, Politics, Education

Sociology, Second Level, (Economy, Law, Ethics, Morale, Aesthetics, Religions, Art, Politics, Education) (cont.)

Finally, let us incorporate Sociology into our chart:

Chart 18

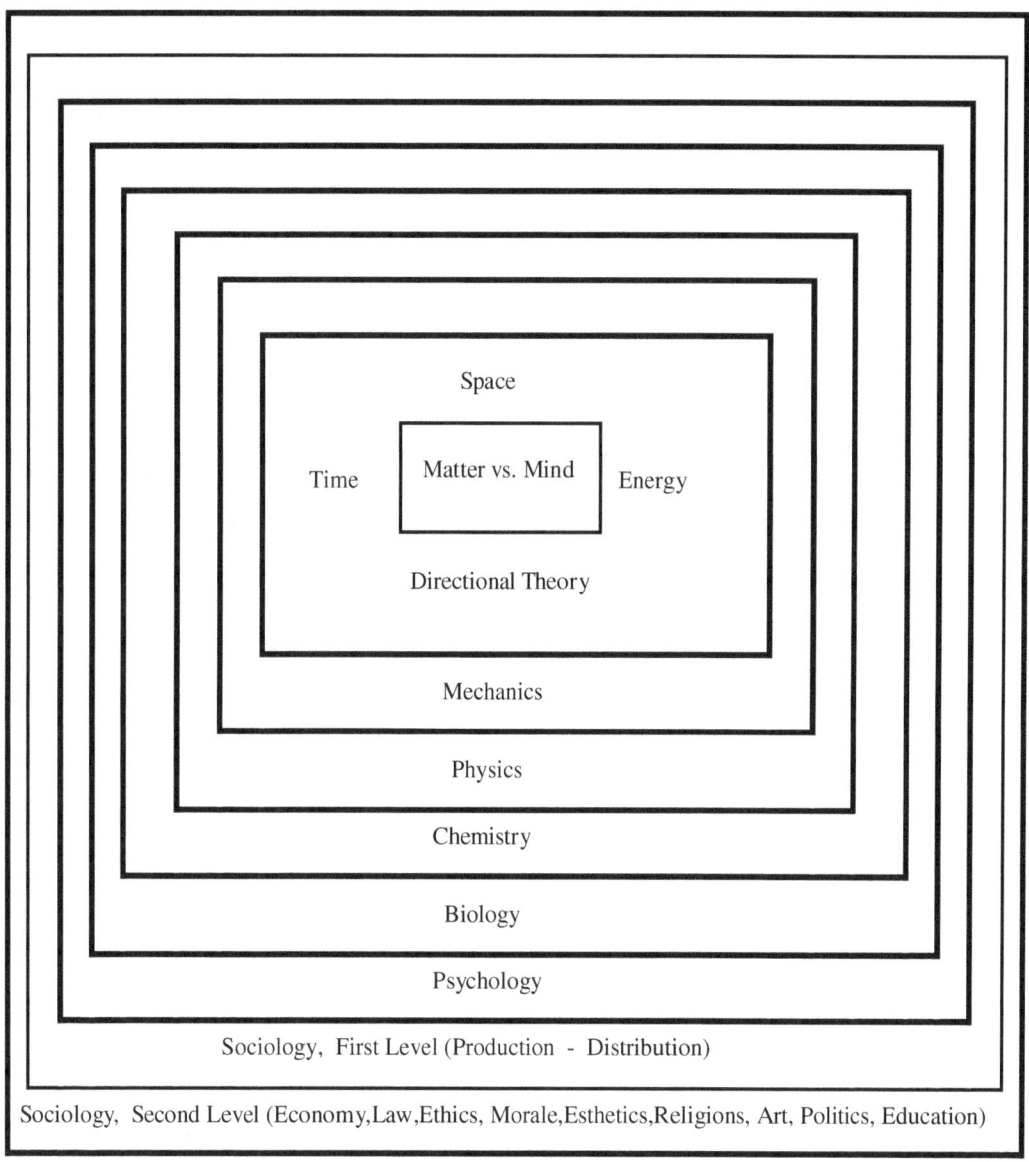

That concludes Chapter 7, so time to move on to the next subject of my Philosophy, that is the Interaction of Intelligent Species Theory.

Part 1 - Chapter 8

Interactions of Intelligent Species Theory

Interactions of Intelligent Species Theory

To complete our system we have to spend a short time with the Interactions of Intelligent Species Theory. This will be the next level of complexity of Matter and Mind. Today in the beginning of 21st century we don't have any conclusive proof of other intelligent existence, only lots of speculation. We continue to encounter other life forms in our planet (animals) like dogs, dolphins, monkeys, that seem to have an "intelligence". If we do a short investigation, we will realize that all they "seem to know" as an "intelligence" is learned through the training that humans provide. I argued in the earlier chapters about the 1, 2, and 3 Levels of the mind (Chart 11). I will repeat myself here, that I am convinced that in the animal world that the Level 3 mind doesn't exist, or it exists in a very early stage of their development.

The real encounter will be when we (humanity) will meet an existence that is capable, like us, to build their own artificial society or societies as humans do on Earth. Of course here I am jumping way ahead of myself, because as I said earlier, we don't have a clue about any such possible existence. The shape and form of their existence is only imagined by us (humanity) so far through our own science fictions. So we don't know if it will be a society or only one individual, if it will be "carbon based" or other, or if it is only a big question mark "???".

My philosophy can give us small directions here to say that "they" will have a combination of matter and mind in some form. They will be subjected to the laws and rules of Mechanics, Physics, Chemistry, Biology, and whatever their Psychology, just like us (humanity). Our encounter, whenever or if it ever happens, will be dangerous business for humanity and also for the other life form. I am hoping we will take a small step in time to "get to know each other" instead of a fast one that can lead to disaster.

I know with this chapter I open myself for a lot of criticism from many angles. It is easy to understand, that when I say we didn't encounter any other intelligence, I declined to accept any current 21st century proof for the "higher power" existence. I have to repeat here that McKaneism doesn't conclude on that issue.

Part 1 - Chapter 8
Interactions of Intelligent Species Theory

Interactions of Intelligent Species Theory (cont.)

Without any empirical evidence, one way or the other, I have to take a stand, that humanity is lacking the knowledge for that decision. Also, in my mind we didn't find "real evidence" for UFO visitors either.

I can't say our governments in our planet have or don't have any evidence, because for obvious reasons they decline to share the knowledge, whatever they have or are lacking. The obvious reason is that they won't be able to control their societies anymore if by any chance the answer is "yes", they have such evidence. Humanity isn't ready for that encounter yet. Just imagine the response, and how far reaching a real "intelligent life form" encounter could be.

I had to incorporate these thoughts into this book, but I have stop here because I don't see any further in the complexity of matter and mind. It is possible that there are more complex layers that exist as a continuation of my chart, but I have to leave that discovery for some other minds. The next page will show a completed chart of the matter and mind complexity as we finished our journey of building, step by step, the chart that I introduced in the beginning of this book. (Chart 2)

Part 1 - Chapter 8
Interactions of Intelligent Species Theory

Interactions of Intelligent Species Theory (cont.)

Finally, let us incorporate the Interactions of Intelligent Species Theory into our chart:

<u>Chart 2</u>

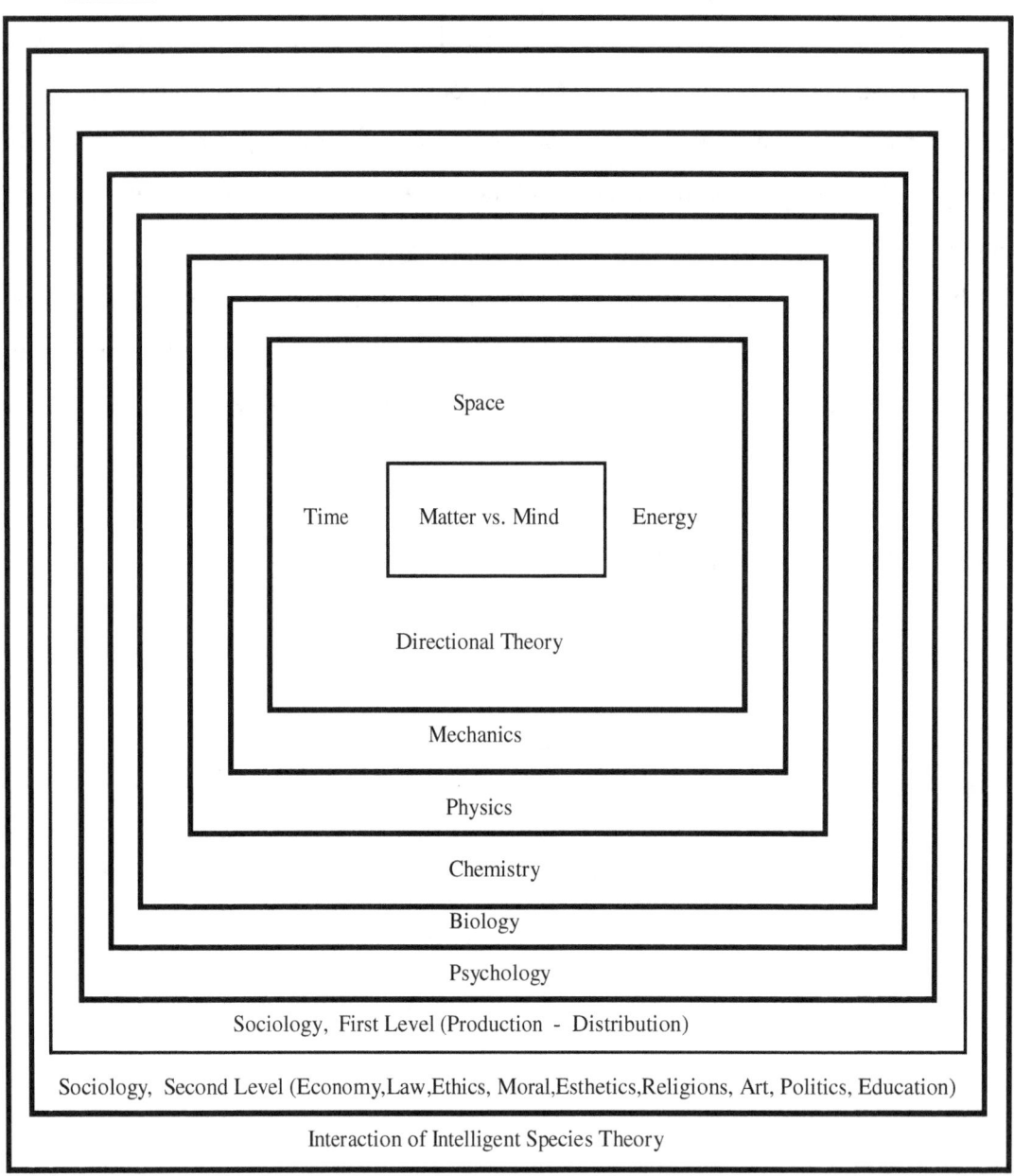

Part 1 - Chapter 9
Solutions for the future (One Language, One Nation, Way of Human Life Theory)

Solutions for the future (One Language, One Nation, Way of Human Life Theory)

Let us try to see into the future in this chapter, so we will dream about it a little. Of course all the Philosophies in the history of Humanity try to explain, analyze the origin and past, direct the present, and foresee and recommend the future. McKaneism isn't different in that sense. Based on the observations and discoveries in this Philosophy I hope I started a new way of thinking and viewing humanity. I hope I did build "The Foundation of Humanity" where the people of the 21st century can understand themselves better. This could give a chance to humanity to progress in the future with more self-understanding and a vision that can lead humanity to a lot better situation than we have today. I know, very strongly, that this is a very solid foundation, and hope many more philosophers and many other people of the current existing human societies will feel the same way. Now let us jump into the vision of the future of McKaneism.

I will investigate and make suggestions in three major areas of the human existence. My conclusions are (in McKaneism) that humanity should organize the existence on planet Earth to One Language, One Nation, and I will conclude and explain the important issues of humanity in the Way of Human Life Theory. (And a promise like before that I will return to some of these issues in more detail in Part 2 in this book).

One Language

We know that people in human societies communicate through languages. The languages themselves are an ever evolving, changing set of sounds, that provide the way for one human to communicate (share the content of his or her mind) with another human.

Throughout the history of humanity many different languages evolved in different regions (continents) of planet Earth. With the technical improvement in human societies, now we are able to travel to every location on Earth, and we are able to communicate in such a way to translate our languages into each other. As I was writing before, I had a great fortune and misfortune to speak two languages. My native language is Hungarian, my learned language is English (American English).

Part 1 - Chapter 9
Solutions for the future (One Language, One Nation, Way of Human Life Theory)

Solutions for the future, One Language (cont.)

I understand a little in Russian, because I learned a little in elementary school, and I understand a little in Italian, because I was lucky to spend a few months in Italy. Through my martial art practices (Judo, Aikido), I picked up a few Japanese words.

I consider English my main language because this is the language that I use in my every day life. However, what I run into time to time are some Hungarian words (that after more than twenty years in the USA) that I still don't know in English without looking them up in the dictionary. The final analysis is that most language (especially the so-called common daily language) use is the same — meaning that all you do is express them with different sounding words in the different languages. It easy to see that most languages have words for "table", "spoon", "chair", etc (so called things). It gets a little more complicated with the emotional description of words, because they may vary in the different languages. When it comes to phrases and concepts, we can run into some trouble. For example: the Chinese word "chi" has no one word translation to English, but we can use many sentences to describe the meaning of the word. A close examination can show that (at least in my knowledge) we can always translate any thoughts, spoken in one language, to another, with more or less effort.

The questions are: Why one language? How can we (humanity) get there?

The next few decades in the life of humanity will decide that either humanity is moving toward a survival or a self-destructing direction. The choices that humanity needs to make (together, everybody included) in this planet will require a great deal of understanding and communication. We have a good example on Earth with the countries like Great Britain, Canada, Australia, and USA that have very little misunderstanding and no fighting between them at all. Interestingly, they all speak the same English language (of course with some dialect differences). The same language makes these countries feel like a big special family. They have from time to time had some great disagreements on issues and hard debates like any other "family", but those never escalate to the level of fighting a war between them (at least not recently). This is a good example that the whole planet can follow.

Part 1 - Chapter 9
Solutions for the future (One Language, One Nation, Way of Human Life Theory)

Solutions for the future, One Language (cont.)

The great advantage of one language is that it can bring the people of this planet very close to each other in understanding. For example, studying philosophy I read the Greeks: Socrates, Plato, and Aristotle in Hungarian and then later in English translation; the German: Kant, Hegel, and Marx in Hungarian and in English translation; the Russian: Lenin, in Hungarian translation; the Bible in Hungarian and in English; excerpts from the religions of Islam, Judaism, Taoism, Buddhism, Hinduism, and Confucianism in English and thereby relying on the words of the translators. That gives anybody the definite disadvantage in understanding the extremely complex minds of those people who lived before me in history and contributed so much to humanity.

It is humanly (I think) impossible for the average human to learn four or five languages in the level and proficiency that is needed to truly understand all these people in their original languages. Learning through translations, I always had many questions, such as did I understand the thoughts well or did the translator do a complete job? Of course my situation isn't the only one. Many scientists and many other people of the international communication area are running into the same problems of translations every day in their life.

I hope I convinced you about the one language necessity, so let us take a look at how can we get there (the way I think it is easiest). First, we take the next generation in every country and in every school, starting at age six, and start teaching them one language (for example: English). (I would prefer English for the language because I found that with 500 to 1000 words, a person can function in an English language society at the basic survival level. Anybody can try to pick up English as a second language; it can be a very quick and easy process to pick up that many words. Usually a six-month part-time English course will do it.)

So, a new beginning can be there. Don't misunderstand me, I have great respect for every nation and every language — growing up in Hungary, I encountered literature, novels, dramas, paintings, poems, etc. from the heritage of more than a thousand years of Hungarian culture.

Part 1 - Chapter 9
Solutions for the future (One Language, One Nation, Way of Human Life Theory)

Solutions for the future, One Language (cont.)

Each would deserve to be introduced outside of the country, however, I've found very little signs (for example in the USA) of that introduction. I can speculate that many other countries are in the same situation. Out of my deep respect for every heritage of every country, I think the following steps should be taken:

During the same time as the new generation is starting a new life beginning to learn one language on Earth, the artifacts of existing cultures, of every nation, should be preserved, digitized, and stored in computers. (Here I come with my own profession, but I strongly acknowledge that books, pictures, the pictures of architecture, historical writings, special blue print designs, etc. all can be digitized and stored, preferably in multiple copies and archives!). At the same time, begin starting to translate all of these artifacts to the chosen language (here for this example, English), and those can be archived also. It easily could take a generation to do that. (Today we think about twenty-five years is one generation within our societies). One more important thing is to establish the study (classes) in the universities of all nations, cultures, and languages (as a second language) so that it can be preserved for the future of humanity. By the end of the second generation, about fifty years, the planet can complete this translation to the new world and this is a "very short time" considering the time of human life or history. I see it in my own son that he, as a second generation in the USA, has no problems with a new (English) language, however, he still has an interest in the Hungarian language and culture — it does not affect him (other than in a positive way) to function in the society of the USA.

All those people who think it can't work must set aside their own **nationalism** (another highly charged emotional mindset in the way of progress) because that has long outlived the interest of humanity. Looking at my own experience, I realize that I don't think any less of my country of origin (Hungary), but I prefer to speak English, and it is much more practical to speak one language than it is being divided. It is just a matter of getting use to it.

Part 1 - Chapter 9
Solutions for the future (One Language, One Nation, Way of Human Life Theory)

Solutions for the future, One Language (cont.), One Nation

Once someone lives by and speaks a new language, then it becomes second nature, and they won't miss out from the start of origin. The other major benefit is that humanity, as a society of one, can become a family of one (humanity). We will be able to understand each other, and each other's problems, and work on the most pressing needs that humanity ignores as of today. This will be one of the most important steps in the survival of the human race.

One Nation

The second most important step is to form a **"One Nation"** society on earth using the model of the **Fair Share Distribution System** for humanity. Humanity, from time to time, gets together in meetings on political issues or sporting events, where somehow the differences of societies seems to disappear. The reality is that, on the Earth in the beginning of 21st century, there are about one hundred seventy (170) nations that exist. The planet Earth is divided by lines (of course only in the human mind) to separate different nations. That is the result of different Production-Distribution systems.

Each society has their leading group of individuals whose life is many times much better in each society than the many other people in the same society. These are the people who are keeping this current system functioning, and these are the people who are protecting the current systems, mostly for their own good. To form a **"One Nation"** society on Earth is far more difficult than the creation of a **"One Language"** society on Earth.

The interest of those so called "leading groups" of existing societies, as of now in the beginning of the 21st century, try to override the interest of humanity. Anyone with a little vision can see that the interest of humanity will win and override the interests of these few small groups, but the road isn't certain yet. Today in those countries' distribution systems the so called "Power" is distributed to those few leaders. They have a power to control. Of course they like it too, that is the reason they are seeking it. If one looks at someone like myself, one can see that I created this publication to search for the best possible existence for humanity.

Part 1 - Chapter 9
Solutions for the future (One Language, One Nation, Way of Human Life Theory)

Solutions for the future, One Nation (cont.)

I am not interested to be a billionaire or be a suppressive leader or be anything of the societies that exist in many countries today. They are all outdated, and unfortunately, are driving humanity in the wrong direction. (if you remember, the "Directional Theory").

The concept of the **"One Nation"** society on earth brings up another big problem. Today, because people of some countries have a better social form of existence, (better constitution, better social structure) and some have more natural resources than others (which by the way those resources actually belong to everybody on Earth), these people accelerate at a faster pace than the others. That is measurable with the standard of living (economical measurement), where people in one country have a better life than people in other countries. The merging to one nation slowly began in the union of the European nations, but it is a very slow and measured union. In other cases, the structures of the societies are not compatible at all. All nations in the European union were capitalist societies, with similar production-distribution systems. Can you imagine the difficulty of the merge between the USA and China?

However, it is the interest of the survival of humanity. The only way I can see this happening is if all the societies on Earth adopt the model of the **Fair Share Distribution System.** This could give each of them the chance to have a similar **Production-Distribution System** in place, then the careful step by step merging process could begin between societies.

Way of Human Life Theory

I called it theory here, but it is more like a recommendation or a possibility or maybe even a necessity for humanity. The **Fair Share Distribution System** provides the fairness needed in society to address even the most complicated issues.

Humanity currently has a Direction into a dead end. Why am I writing that? With the current exhaustion of resources on Earth combined with the current overpopulating effect and over pollution effects on Earth, it is getting into the level of a disaster.

Part 1 - Chapter 9
Solutions for the future (One Language, One Nation, Way of Human Life Theory)

Solutions for the future, Way of Human Life Theory (cont.)

Because (as we argued before) the planet Earth is a closed system, it has a certain capacity to support life and sustain life in the continuation of a life cycle. We are already making our big mistake by overpopulating this planet. This all can't be solved at the countries level, because each country, by national pride, is looking at the other countries to cut back on their consumption, pollution, and population. Of course, no one is out there today with a positive example. Today it is unthinkable to say to the people of Earth that the population voluntarily has to be reduced to 2.5 billion from 6 billion. Also it is unthinkable by some people today, to tell a rich individual that his or her wealth needs to be distributed differently.

From country to country, and from group to group, within each country we have a major problem coming from the Distribution systems of all societies. We pointed out before that the Communist system won't work and we still have nations with many millions of people who are living that social format. The Capitalist system won't work either, but tell that to someone in the USA or Germany, then they have the first response that "it worked in the previous centuries, why not forever?". Most people do not actually realize that it is not working. We've exploited the natural resources on Earth, we've created countries where some people have billions, and yet millions are homeless and flat broke. That does not work in the long run. We are encountering more and more problems, but nobody in the leadership of many countries wants to address them.

They all are ignoring the fact that with the current pace of society, we will exhaust our natural resources (especially oil, that takes millions of years to form underground), and end up in an impossible to handle situation. The time is *now* to address these issues (already a little late, but waiting any longer will make the situation even more unmanageable).

Part 1 - Chapter 9
Solutions for the future (One Language, One Nation, Way of Human Life Theory)

Solutions for the future, Way of Human Life Theory (cont.)

Here are the few changes that must take place:

(It isn't a recommendation, it is a necessity in my view!)

1. The world's population must be reduced to 2.5 billion. (voluntarily!)

2. The world hunger has to stop by introducing and adopting the "Fair Share Distribution System" in every country.

3. Change the world into a "One Language" society.

4. Change the world into a "One Nation" society.

5. Humanity must learn from nature to make every manmade product 100% recyclable as things are in nature — the 0 pollution system.

6. Start creating Earth-like bases for human existence on the other planets in our Solar system.

7. Reduce the Nuclear weapon and pollution problem with "smart agreements"

8. Start gaining total control over diseases. (Vaccination, Medicines to stop them!)

9. Stop Chemical and Biological weapon production.

10. Develop a progressive "Value System" for humanity to live by. (see more in Part 2 in this book).

These are just a few of the big problems and solutions to start with. The sooner humanity wakes up the better and then the sooner we can see some serious results. For all those people in politics who are always referring to their children and to creating a better place for them, it is time to wake up and get to work. The time is running out for humanity. We are facing the biggest problems in this planet and we are not ready to live in space yet on a permanent basis. We are ignoring the devastating possibilities of the "man made" (nuclear, biological weapons) and "nature made" destructive forces, like "large meteorite strikes" on Earth, that can eliminate the existence of humanity. (Of course a lot more coming in these subjects in Part 2!)

Part 1 - Chapter 10
Philosophy, The organization chart of Philosophies

Philosophy, The organization chart of Philosophies

In this chapter we will investigate the different Philosophies that have existed previously on planet Earth. My intention isn't to analyze or compare all the previous "Systems of Thought" or "Philosophies" word by word, because first of all I don't have (and don't think any human being has) that detailed of knowledge of all of them and second of all, it can be a subject matter for a many volume library to write something like that. However, all "Systems of Thought" and "Philosophies" can be represented in an organization chart, based on where they belong on the basic question of Matter vs. Mind. Reading them I realized that all of them try to answer this urgent question of humanity from the beginning of human time. Humanity always (even today) has many unanswered questions, and they call for a "System of Thought" or "Philosophy" to give it an answer.

In my next chart I will represent only the major Philosophies, but one can extend this chart easily, after understanding the concept of building it. The concept is very simple. Those Philosophies that conclude that "Mind over Matter" is column one. The ones that don't decide is column two, and those that conclude "Matter over Mind" are in column three. In naming or labeling them I will use the same wording that was approved and used by humanity throughout the history of humanity.

Here is our chart:

Chart 19

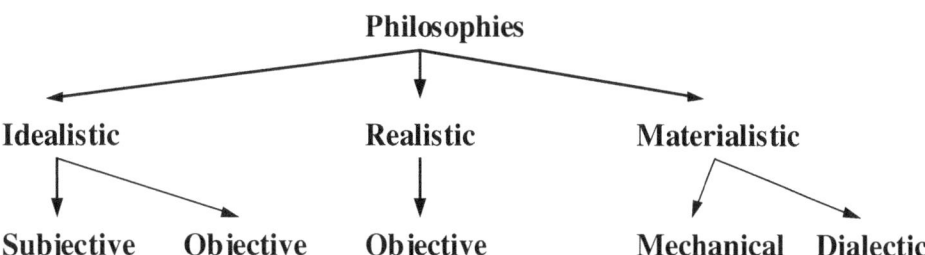

These are the names that humanity used throughout human history to name and group their "Systems of Thought" and "Philosophies".

Part 1 - Chapter 10
Philosophy, The organization chart of Philosophies

Philosophy, The organization chart of Philosophies (cont.)

Going further, I will give some examples with short explanations of each to represent their main line of thought process. It won't be detailed, but I hope it will give an idea of differentiation between them.

To analyze closer and deeper, these "Systems of Thought" and "Philosophies" can take one a life time of reading and understanding in many different languages, like Greek, Latin, English, German, Russian, Chinese, Japanese, and the languages of India just to name a few of them. The additional difficulty is that some of these languages have many dialects, or the "Systems of Thought" and "Philosophies" were written hundreds or thousands of years ago, where the languages don't exist exactly in the current format or in the way as we have our modern languages today. So, overcoming these challenges (I hope humanity will one day) is not a simple task. For now, let us just focus on some examples of those categories for our better understanding. Here are a few examples (these are not intended to be complete) to help our understanding of these categories.

Chart 20

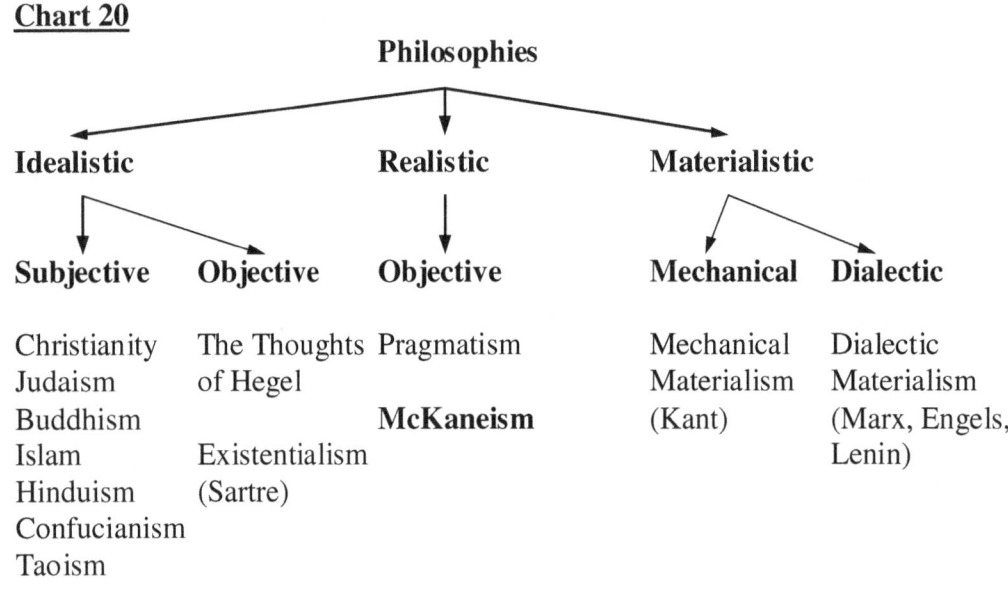

Part 1 - Chapter 10
Philosophy, The organization chart of Philosophies

Philosophy, The organization chart of Philosophies (cont.)

These are only a few examples, anyone can find more throughout human history. Sometimes it is difficult to differentiate where some of these "Philosophies" really belong. For example, Confucianism is one that may belong to the Realistic philosophies, but basically the belief system of "Unity of Heaven, Earth, Human" makes it more Idealistic than realistic.

Of course, we may need to investigate his idea of "heaven"! Pragmatism isn't a Philosophy as far as I see it, more like a "System of Thoughts" where the focus is based on the "Cash Value" of things. In the USA, we can relate to it easy, just think about how often we hear the sentence "I can or can't afford that." However, pragmatism can't give any explanation over things (or give a wrong one) where cash value can't be or shouldn't be placed. (For example: life, love, morale, law, health). Unfortunately we (humanity) place a "cash value" on those from time to time.

I will spend more time in Part 2 of this book to investigate and analyze some of the major elements of those "System of Thoughts" and "Philosophies." After all, humanity was and is being influenced by those, throughout human history, including in our modern time. Most people have one or sometimes more than one of those philosophies mixed inside of their way of living or thinking.

Many times some individuals don't even realize it, where he or she belongs. The understanding is very important, however, because these philosophies determine the thinking and behavioral way of life for those who understand and practice them. Many times in human history we had a collision of philosophies that unfortunately lead to a collision of humanity. When large groups of people (countries) can't debate their thoughts or can't find a solution for their differences, many times humanity ends up sorting out the problem in a violent way through terror or wars.

Frank P. McKane

Part 2

Frank P. McKane

Part 2 - Introduction
McKaneism - The Foundation of Humanity

Introduction - McKaneism - The Foundation of Humanity

Part 2 Introduction, Why ? One can notice that our kids in elementary school don't start learning calculus. In the human development, first we start learning the numbers, then we learn more complex mathematics in junior high school, then more in high school (some of them), and then more in college or a university — where we are finally introduced to calculus. I will try to follow the same concept with the introduction of my philosophy, McKaneism. First, I made my case on the statement base and built a logical, reason based "thought system" in Part 1, with no other explanation than it could make sense to anyone who is willing to think it over in one's own mind. Now it is time to enter a "college or university" in Part 2, so I will get more serious and introduce the "same thought system" as Part 1, only with more complex reasoning and fact based McKaneism with explanations and conclusions.

In Part 2 I will push my Philosophy to the limit. I will introduce all my observations, reasoning, thought processes, and conclusions. Some of them will be way ahead of the way humanity understands itself today. Of course that will create the conflict between my thoughts and the thoughts of the current "World". In my Philosophy, McKaneism is the first time in human history where I will introduce the way of the Level 3 thinking (Chart 11). That is the Logical/Reasoning way to look at the world around us. This will be new to the current Level 2 emotions based world. Because of the relativity of my philosophy I established a system in Part 1 that describes (in my view) the known universe around us in the year 2003. Notice here I am saying "Known Universe", that means I never intended to look into a future for thousands or millions of years and tell you what will happen and how we (humanity) will exist there and what will be our knowledge base there.

In Part 2 I will go all out, no holding back! That means that Part 2 is intended for readers who are deeply studying philosophy or for those who are turned to philosophy in their sciences, like biology, sociology, etc.

These are the scientists in today's 2003 societies whose views are suppressed most of the time in their current societies.

Part 2 - Introduction
McKaneism - The Foundation of Humanity

Introduction - McKaneism - The Foundation of Humanity (cont.)

Just think about an anthropologist who has investigated a 5 million year old bone remain (in secret of course, because "naturally" that Darwin was "stupid", and "God" already took care of the creation of the world, and therefore the case (and minds) closed!) or a biologist whom decoded the human DNA (one of the greatest achievement of mankind!, and now they try to use this fantastic knowledge in medicine for mankind and mankind saying that they are "breaking the man's law?!") — and these people are hunted by very narrow minded politicians and (child abusing) clerical losers (like the witch hunt in 1200).

Of course these strong words don't apply to all, only to the many guilty ones!, who are looking at this as a "political power game", or a "who gets the big bucks income game" or a where is the governing and ruling fate game". They don't realize that if our smart scientists are   correct and we ignore them **it can be the end of humanity!** ("Mega Tsunamis", "Mega Volcanoes", "The airborne X bacteria", "The piece of rock with the 10 miles diameter, that is on it's way with one third of the speed of light." — forces to be matched by Level 3 based science, and not Level 2 based emotional fate!)

It is the most serious thoughts that you have ever read in your life. Make no mistake about it, I use strong words because this subject is "deadly" serious. Humanity has a choice, to take on the challenges of survival, or do nothing and become extinct as the dinosaurs.

If you think "big deal, Heaven here I come", I will address that thought and all those previous statements and observations for you later! The question isn't that "can all those events really happen?" but instead the question is "when will any or (worst case scenario) all of them happen"? One can study those sciences and the probability theories in mathematics and can see for oneself the validity and reality of the arguments. Reality is the philosophy that McKaneism is based upon (in those previous possible events) — the "world of physics, mathematics, biology" has nothing to do with our emotions like fear, "feels good", or "don't worry God will save us". Those emotions may guide some of us (humanity) in the year 2003 in our average daily life, but they won't save us (humanity) from the reality of the laws of physics, mathematics, and/or biology.

Part 2 - Introduction
McKaneism - The Foundation of Humanity

Introduction - McKaneism - The Foundation of Humanity (cont.)

Facing the events of those realities we (humanity) need to use our Level 3 Logical/Reasoning capabilities, to survive and have a prosperous future. Of course one will see many arguments later in this book, that we (humanity) never can, and should not throw away, our emotions because those are part of our human existence. However, dealing with reality, we need to stand on the reality base, and control some of our emotions (discipline them) like love, hate, happiness, sadness, etc. The reality of those possible catastrophic events are fact based, and the only way we can comprehend and counter them is if we use our Level 3 Logical/Reasoning thinking.

If you think I am saying that it is time to get scared, that won't reflect the truth either. That is just another emotion. Time to get our (humanity) knowledge together and try to face reality, that is exactly what I am saying. Part 2 was written for those who are knowledgeable in the philosophy or "thought processes" of our reality driven predecessors.

In Part 2 - Recommended Readings, I will list a long list of publications that one needs to read, digest, and understand before one can understand Part 2 of this book. The reason I give my readers a fair warning is because without the deep understanding of those predecessors in Philosophy as a science, one won't have a chance to understand Part 2 of this book. One needs to understand that all of those predecessors wrote a small half library size of volume of books. It is impossible here for me to reiterate or quote all those works in Part 2 of this book, so I will refer to them as investigated, criticized, and concluded knowledge of their work (in my mind), whereby some are approved by the logical thinking of McKaneism, while some are rejected, based on their merit or lack of it.

If you disagree with me, "Welcome" to a thirty years part time reading and study ahead of you before we can connect to argue (you will see this for your sadness on the 3D Knowledge Diagram). Looking at the fact that I am an average nobody and some of you are the "smart guy" (you figure out how much smarter you think you are), that will cut down on your reading. For example, if one of you is twice as smart as I am, you have only fifteen years of part time reading ahead of you, etc. If one thinks it is playing with words, one needs to check the concept of the "Knowledge Diagram" of the individual human being.

107

Part 2 - Introduction
McKaneism - The Foundation of Humanity

Introduction - McKaneism - The Foundation of Humanity (cont.)

One can realize that without **some overlapping knowledge** it is impossible to debate any subject. For example, one can tell me all the arguments about the college basketball games, and one will find out fast that my knowledge isn't deep enough to argue on any of the events. It isn't that I am not interested in college basketball, but not attending four years of college in the USA, I thereby left out four years of my life when people focusing on college curriculums and of course their basketball actions.

I will extend in the later chapter, on **overlapping knowledge,** because that is the key to human understanding and communication.

One more observation, that Part 1 of my book was designed for "beginners" and Part 2 for "advanced" readers. I'll let you decide how "beginner" are "my beginners" who have read, understand, and enjoyed Part 1 of this book of science of philosophy! (Sorry, not a 5th grade level of reading!) We (humanity) may have to design a better education system to insure that we don't have a situation where 25% of the nation (like USA, and it is good on this planet Earth!) is on the 5th grade level of reading. Many times I used for example, the USA, for better understanding of some of my statements, without giving any lengthy explanations about it. One needs to understand that I use one of the most technically and humanly advanced societies for my arguments. My reason here is to wake up humanity to recognize in other countries, "where" are they in reality, if I can find so many questionable issues in one of the most advanced countries (USA).

In my new philosophy, McKaneism, I will introduce a new concept, a new way of looking at our existence, as humanity, in the more realistic and more logical way. I will also investigate our current "World" (Earth) of the 21st century, and the past, including many very popular and unpopular thoughts, and some future possibilities, as far as one can venture into speculation in the future without losing site of reality (using the method to investigate the current trends, and extend on them to forecast where those trends lead our humanity). I will do all this within a framework of my new philosophy McKaneism, that will provide answers and explanations for most of the questions or at least will look at them in a realistic way.

Part 2 - Introduction
McKaneism - The Foundation of Humanity

Introduction - McKaneism - The Foundation of Humanity (cont.)

I have to repeat that I have no intention claiming that I will provide "all" the answers for humanity, because it is an impossible statement and impossible action to try with the knowledge of humanity in the year 2003 (more than that I have only a subset (small subset) of the knowledge of the 21st century humanity).

However, I strongly know that my observations, reasoning, and conclusions will induce the thinking of many people, that will find these investigations and arguments valid, and are willing to open their mind to look at our universe in a different spotlight (I am definitely counting on those, who against all odds, being raised in the 95% fate based society, are still willing to open their mind and listen for reasoning and thinking arguments).

I will extend Philosophy as a science to connect and guide the other sciences in Part 2, that is one of the most important features of McKaneism. My observations, reasoning, and solid logical conclusions will show that only philosophy can explain the way all human knowledge is connected, and lead to the universal explanation of the questions of humanity and intelligent existence. The other intention of mine to organize human knowledge and also place the current emotional based societies in the right place with all those emotions.

I never intended to construct "The Foundation of Humanity" without emotions. Emotions and the Level 2 emotional system is part of our existence. That is what we (humanity) are! We are a Level 1, 2, and 3 based existence (Chart 11). The argument here, that "The Foundation of Humanity" has to be built on the Level 3 Logical / Reasoning base, then we can apply the Level 2 and Level 1 for the rest of the building — as I put it previously in Chapter 6, **"The Foundation of Humanity" + "The rest of the Human Castle"**.

For example, the creation of this book is coming from a very emotional influence. For many years I was interested and studied Philosophy, and to see how lost Humanity is in the 21st century, and the compassionate view against the thought that the few "Stupid Ones" can get the power in our current societies and that they can "terrorize" the rest of the society.

Part 2 - Introduction
McKaneism - The Foundation of Humanity

Introduction - McKaneism - The Foundation of Humanity (cont.)

If one disagrees just turn on your daily news and check it out — how many countries have small "warlords" and "dictators", who live like "Kings", and in the mean time the USA and the advanced industrialized nations must deliver many a million dollars worth of food to the suppressed, hungry, and starving people of those countries.

Is this a proper set up or design for an "Intelligent?!" life form? I'll let you answer that question!

The other issue is that one can look into all our religions, and find that people in the name of the "peaceful religions of Judaism and Islam" are bombing and killing each other every day. Of course the "good old play off between Christianity and Islam" is another "not so good!" story either. All these **"peaceful?"** religions are ready to wipe off the others from the face of this planet if they could have the proper military power to do it.

Why? In the name of their own personalized God, the one and only correct one? The answer is simple, their emotion based, many thousands of years old belief system is outdated, and it doesn't support the needs of humanity anymore. The basic production-distribution structure that none of these thought systems address makes the emotionally charged humans hate each other and hate each other's religion.

One needs to understand that these few mentioned religions aren't the only problem — the structure of some kind of "communism" in some countries, the modern European existentialism mixed with pragmatism, and the One religion + Existentialism + Pragmatism based USA structure won't provide the proper solution for the human existence either. None of these philosophies are based on the "Attributes of Human Being Theory", and because of that they don't support the needs of humanity.

The reason I have to have the disagreement with them is because they all provide a "screen door" for an individual to look at the world. The wrong screen door can distort one's visions and reality. I will spend a lot more time to analyze the pros and cons of all these philosophies and beliefs in Chapter 17. For now I just need to conclude based on all these observations that it is time for a change.

Part 2 - Introduction
McKaneism - The Foundation of Humanity

Introduction - McKaneism - The Foundation of Humanity (cont.)

It is time to introduce something way more advanced, way more appropriate Level 3 based structure that humanity can live by and survive by in the millenniums to come. Yes, you guessed it right, that is the **Philosophy of McKaneism**. One thought process that works for all humans whom willing to think.

Observe, reason, and conclude on your own, and find out for yourself why McKaneism is the Philosophy for the future — the Philosophy that provides realistic explanations for the way all human knowledge is connected and leads to the universal explanation of the questions of humanity and intelligent existence.

What is McKaneism ?

McKaneism is a philosophy, a system of thoughts and thought processes, that is based on logical and realistic observations, thinking, and conclusions — a philosophy that establishes "The Foundation of Humanity".

The main strength of McKaneism is the flexibility of this philosophy. The definition of flexibility: It doesn't mean that if you say green I say green or if you say blue I say blue. The **structure of complexity of Matter and Mind** that I described in Part 1 is coming from observations, so I am not flexible to change it without a very compelling argument that proves me incorrect in my observations. The flexibility of McKaneism is that this philosophy does not conclude or decide on thoughts that are non-factual or thoughts that can't be proven by one of those sciences at the area of observation.

Unknown human knowledge will stay unknown (until a true discovery later in life), and I will analyze it as hypothetical without final conclusions nor take it as facts.

This approach can keep this philosophy flexible and open minded. McKaneism will grow with humanity, based on the Directional Theory.

Part 2 - Introduction
McKaneism - The Foundation of Humanity

Introduction - McKaneism - The Foundation of Humanity (cont.)

What are the major areas of study of the philosophy of McKaneism ?

Basics	-	Fundamentals as Matter vs. Mind, Space, Time, Energy, **Directional Theory**.
Physics	-	the new look at the matter and the theory of the Big Bang.
Chemistry	-	the look at the special element that has the capability to form a great variety of chemical components.
Biology	-	**Attributes of Living Entities Theory**.
Physiology	-	**Knowledge Theory and Chart extended to "3D + Time Universe", Attributes of Human Beings Theory**.
Sociology	-	**First Level, Fair Share Distribution Systems Theory Second Level and Interaction of Intelligent Species Theory**
Future	-	**Solutions and possibilities for the Future of Humanity**.
Philosophy	-	**McKaneism as a philosophy, Realistic Logicism, Value Systems of Societies, Philosophy in the World of Sciences**.

The **Bold** letters on the right above represent thoughts and thought processes that are new as a philosophy of McKaneism, because they are new as an observation or because they are new as conclusions of my observations.

One more important comment: Part 2 is not structured exactly as Part 1 as far as the chapters, because I have more to extend on one subject than the others, however, the "Logical structure" is the same, even if the chapter numbers don't correspond! One can realize that any knowledge of humanity is based on the previous knowledge of humanity.

Let us start to venture into Part 2 and discover the "extended" deepness of McKaneism. I hope one will have an unprecedented logical, historical, and philosophical journey of one's life. I asked you many times, and I am doing it again, don't believe me, *journey* with me and with my thought processes, and draw your own conclusions.

Here we go: McKaneism Advanced as I promised!

Part 2 - Chapter 11
Fundamentals, Matter vs. Mind, Space, Time, Energy, Directional Theory

Fundamentals, Matter vs. Mind, Space, Time, Energy, Directional Theory

We all know from our practical experiences in life, that to build a system (like a house) we need building blocks (like bricks). To build the system of thoughts the same principles apply. Any philosophies that my predecessors in philosophy built, started with the basic building blocks of their thought process. This is the most dangerous time in philosophy, because these building blocks will define the thought process for rest of the arguments. That is where many philosophies are closing the door for open thinking.

The most important difference between McKaneism and other philosophies, that McKaneism is a philosophy that is "man enough" to say "I don't know!". We (humanity) don't like the unknown, we need the base, the foundation that reflects our own individual life. That is the "pitfall" of all previous philosophies. They all try to offer a **birth to death** complete system, where one can "**Feel Comfortable**"! I have to break the bad news for all my readers in Part 2, that after you finish reading Part 2 you will feel just about every possible human emotion about me and my book, but you won't be comfortable. **Philosophy isn't the science of comfort,** (as of today, philosophy is not even viewed as a science at all!) "**it is the science of the sciences**". If one wants to feel comfortable, then one needs to talk to a psychologist in a therapy session or one needs to get a massage.

This time in history McKaneism will establish "Philosophy as a Science of the Sciences". Let me begin with basics, knowing that I will put down the foundation for the most realistic philosophy that ever existed in human history up to today. Once upon a time Albert Einstein had a thought that somehow the Universe can be described with a "General Gravitational Relativity Theory".

The interesting thoughts behind it, that this genius man had a deep thought in his conscious and unconscious mind that the Universe (The "Universe" around us that we "see" with our telescopes and radio telescopes, but not always understand), is somehow governed and driven by one big set of "general" rules. He tried to prove his theories within the boundaries of physics and ended up not finishing his work, leaving the world puzzled until someone can pick up where he left it.

Part2 - Chapter 11
Fundamentals, Matter vs. Mind, Space, Time, Energy, Directional Theory

Fundamentals, Matter vs. Mind, Space, Time, Energy,

Directional Theory (cont.)

Make no mistake about it I am not that someone. However, reading and studying his works (and many other publications) I came to the conclusion that physics isn't the science that will describe for us the unified Universe.

I think Philosophy is the science that will be able to provide for us those answers. Throughout the history of mankind many scientists worked with the Matter, Mind, Space, Time, Energy, and tried to observe, analyze, investigate, conclude and describe the nature and underlying principles of those areas.   Interestingly, when ever they run into some of their road blocks, they turned to philosophy. Almost with no exceptions the philosophers of the past also had great achievements in the area of Sciences and/or Art.. (For example **Kant** (Metaphysics) as philosophy, Scientists in Mathematics and Physics, **Sartre** (Existentialism) as philosophy, in Art as a Writer of Drama).

One more observation before I start with the fundamentals. From Part 1 of this book one may remember the definition that our mind reflects the matter as a mirror reflects ourselves. I am very confident that understanding that statement opens up new possibilities for humanity to analyze the way we store the information and reflect back our environment in our mind.

Let me start with a definition. **All the knowledge that we (humanity) have in our mind is also observable in reality or can be discovered or created in reality!**

This is a very powerful, and very important statement, with far reaching consequences. The observation of "Matter and Mind" (in case they both exist) in the relation of "Perfect Mirroring"! I am saying, in other words, if an "Imaginary One" (remember: this is a thought experiment) can combine all the knowledge from all human heads, that would reflect all the matter as my above definition: **observable in reality or can be discovered or created in reality!** To see why it is so powerful a statement, I have a few examples.

Many of you had a lucky situation, to afford to build your own house. So step number one you don't have "your" house in your head, but you have "tons" of examples in your head.

Part2 - Chapter 11
Fundamentals, Matter vs. Mind, Space, Time, Energy, Directional Theory

Fundamentals, Matter vs. Mind, Space, Time, Energy,

Directional Theory (cont.)

It is exciting times (emotionally and of course financially) so one starts to visit all their friends and relatives and starts looking into their houses as samples. One tries to find a good idea in kitchen design or a "slick" gardening setting, where one can sit in their own patio and feel great looking at the perfectly landscaped backyard. Then the builders arrive and flood one's head with blueprints and financial estimates. (Starting to see the point?!) The definition for this situation: **All the knowledge** (about the house) **that we (humanity) have in our mind** (here one individual out of humanity) **is also observable in reality** (ideas from the houses of friends) **or can be discovered** (blueprint from the builders) **or created** (one's final design in one's head first, then on papers about the final house of that one) **in reality!** Very powerful example (**from reality**!) to see how the matter and mind mirror each other.

A few conclusions from this observation. It is impossible to conclude that matter vs. mind, which is first. One can see countless arguments that the creative mind with the images of a non-existing house can prove "mind first", or counter this argument that the creative mind already exists on the physical existence of the matter of the brain, so "matter first", and on and on. The starting point can be picked for the advantage of any side.

For McKaneism I will say the magic words — "I don't know!". In other words, it is unimportant to make that decision, and I will argue extensively in Chapter 17 that this decision creates a non-working closed minded philosophy, and sets a one sided direction for it. The most important fact in this observation is the fact of the "Mirroring between Matter and Mind". In reality a process in time took place before the final house was a mirror image in the mind and then can be built in reality. This is a very important observation; keep it in mind for our conclusions later, but for now conclusion two.

Once upon a time in history, the scientist Copernicus was living in the Castle of the Austrian King. He was a Mathematician with a capital "M", one of the brightest minds in human history, but definitely in his time.

Part2 - Chapter 11
Fundamentals, Matter vs. Mind, Space, Time, Energy, Directional Theory

Fundamentals, Matter vs. Mind, Space, Time, Energy,

Directional Theory (cont.)

He could not practice mathematics, because nobody was interested for that, however, because he also was an excellent astronomer, he was hired to gaze at the stars and get well paid for creating the horoscopes every day for the king and the royal family members. Way to go Copernicus, nothing wrong with that, one can say in "America" (USA), honest good bucks for honest horoscopes.

However, one can see the sad resemblance with our today's world - a brilliant scientist doing his science in hiding, (he had better because the best is just yet to come!) just like our modern anthropologist or DNA biologist (remember Part 2 introduction), or else he gets it from the powers of the "inquisition". (Great stuff! Only 300 to 400 years went by!!! Can we (humanity) speed up a little, before the "big rock" gets here and sends us all chasing after the dinosaurs?!) Here comes the best and most important part in this example:

One day as he was doing his part time mathematics, he discovered that the "math" wouldn't work if the Earth is the center of the "that time known Universe". **In his "math" on paper (in his mind!) he discovered** that the Sun has to be in the center to make all the "math" work. He was in this "marvelous" position that nobody in his surroundings cared anything about his math, nor anyone could understand it. So he sent out the messengers to the fellow mathematicians in his time to check his stuff, because (as any good scientist) first he had doubt about his own discovery.

That traveling discovery put the medieval Europe in great turmoil. (Few people know that one of his followers, a young and very enthusiastic mathematician Bruno Giordano, a bright priest in the Vatican, started to tell everyone in the priesthood that he checked the "math" and Copernicus is correct in his discovery. Because as a scientist, he didn't revoke his words, later he was burned on the stake — way to go humanity: Inquisition one, Humanity zero!) With the technical discovery of the telescope, (of course an early version kind) Galileo observed the planets, especially the four largest moons of Jupiter, circling around Jupiter, and that was the first observation to prove Copernicus "math" was correct.

Part2 - Chapter 11
Fundamentals, Matter vs. Mind, Space, Time, Energy, Directional Theory

Fundamentals, Matter vs. Mind, Space, Time, Energy,

Directional Theory (cont.)

Later he revoked his "own observation" (maybe saying: "I can't believe my eyes?!") before the score could reach Inquisition two, Humanity zero! He decided that he would rather stick around a few more years exiled from friends, house arrest (1633), but alive. (more than 350 years later the Vatican issued a statement that Galileo was a good Christian, but they still didn't say that he was correct!) I read the history of that time in Hungarian and English publications so I am assuming that the story here is exactly correct. As far as the events, they all happened, but one may need to check the timeline of the events. If one is interested, keep reading the corresponding history and science books. One can say, it was an entertaining story, but What the heck does this have to do with the Basics and the Foundation of McKaneism? Good question. One can see from this short story that the **second time we observed** this planetary phenomenon, in this time in human history, that **Matter and Mind Mirrors each other**! Remember Copernicus figured out the structure of our Solar system in the language of mathematics in a time when the whole planet believed that Earth was the center of our known universe. Later, with observation, he was proved to be correct.

The point that I try to make is that leaving the Matter vs. Mind question open can lead to a very open minded philosophy. More than that we need to discover a lot more of the functioning mind to understand "how the mind can have information, then later, can be proved existing in the real world or the real world of science"? The next introduction will be a quick look at the way I define "subset". It is important to understand, when I say that Mechanics is a "Subset" of Physics; I mean that Mechanics with all the principles, rules, and "whatever" of Mechanics is included, embedded, and contained in Physics.

That also means that all the information and knowledge of humanity of Mechanics as a science or topic is included, embedded, and contained in the knowledge of humanity of Physics.

Part2 - Chapter 11
Fundamentals, Matter vs. Mind, Space, Time, Energy, Directional Theory

Fundamentals, Matter vs. Mind, Space, Time, Energy,

Directional Theory (cont.)

Here is a two dimensional chart to represent what am I saying and thinking:

Chart 21

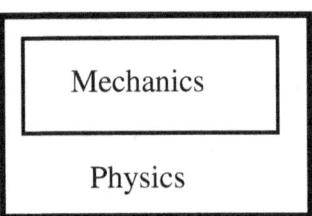

One can realize reading my book that I focus on saying "two dimensional" chart. That isn't a play on words, that has a great implication soon, as I start to investigate and describe the "Universe" around us as McKaneism as a Philosophy sees it. A few more interesting observations. Sitting in the "math lab" in my university years, a theory begins about an 8 dimensional vector area. Some of the advanced students started to describe the characteristics of the 8 dimensional vector area (of course losing me half the time), however, the basic concept that each point in the 8 dimensions has 8 coordinates. (x1, x2, x3, x4, x5, x6, x7, x8) represents a well defined location point in the 8 dimensional area. One can see that here the mathematics describe "something" that we can't experience in our real world. That raises a few very important questions like: Can we have an 8 dimensional existence in our universe? Can it be the same scenario like Copernicus, like the mind discovers something before we can experience it in reality? If 8 dimensions can exist, how about 9, 10, or N dimensions? We know and experience 3 dimensions every day in our life, so can there be more?

If one thinks, this is it, we are finally getting interesting, I have to disappoint that one. I can't answer all those questions in the factual base, however, I will incorporate my hypothetical, logical, and reasoning answers in this chapter.

The importance of this investigation is great because it is one of our fundamentals, that is Space. The "container" we are living in.

Part2 - Chapter 11
Fundamentals, Matter vs. Mind, Space, Time, Energy, Directional Theory

Fundamentals, Matter vs. Mind, Space, Time, Energy,

Directional Theory (cont.)

I didn't find any philosophy, (however I don't claim that I read all the philosophical publications) that addresses this question within the framework of philosophy. Mostly these issues are left on the shoulder of mathematics and physics to explain. That proves my point that none of the previous philosophies can stand up to the definition that they are the "science of the sciences" by not inspiring and interacting with all of them.

My science is computer science; that contributed greatly to understand, organize, and comprehend systems. The many years of working in the information science, working with different computer systems, helped me develop my Level 3 realistic Logical / Reasoning mind. Working with the logical computer science, programs, and systems, for many years also helped me realize that "Nature" is a system, "Society" is a system, just like a "computer operating system" is a system. Looking at those systems closer helped me discover that our society is just as man made and artificial (of course with some natural touch) as a Payroll or Order Entry computer system, of course with a "great deal of more complexity" on the society side. That lead me to understand the structure and complexity of Matter and Mind and how they can form more and more complex structures, as one saw in Chart 2.

Now in Part 2 it is time to extend my observations and move from the two dimensional diagrams of Part 1 to the 3 and more dimensional ones.

We (humanity) as I observed in the practical reality, are living in a three dimensional existence. (The reason I said practical, because in the every day life, we don't address our space and the possible form of our space.) It is easy to comprehend in daily life, thinking of an object like a table, that has three dimensions (I will refer to this later in short as 3D) — length, width and height.

For the mathematical mind that is the axis of the x, y, z, coordinates in the science of geometry. Each point in this system can be represented with the (x1, y1, z1) coordinate triplet. This description of our 3D space coming from Euclid, a Greek mathematician, who is the father of the classic geometry. Declaring the simple axiom that through one point only one parallel line can exist with a line, founded his system of geometry.

Part2 - Chapter 11
Fundamentals, Matter vs. Mind, Space, Time, Energy, Directional Theory

Fundamentals, Matter vs. Mind, Space, Time, Energy,
Directional Theory (cont.)

Later, other scientists argued that one can have many more parallel lines with one line through one point. They established their "parabolic geometry". One can ask: What the heck does all this have to do with philosophy?

The answer is far reaching. One can read the philosophies of Christianity, Judaism, Islam, Confucius, Taoism, Marxism/Leninism, Existentialism, etc. and they won't find any investigation in this level about our surrounding environment that we are living in. They only analyze what is "comfortable for them" or "focused in their view point on things that make their case", and remain quiet about many other issues, like how is the space structured around us that we are living in. One more time proving my point that none of the previous philosophies can stand up to the definition that they are the "science of the sciences" because they do not inspire and interact with all of them.

One can realize that using the three coordinate system, I can build a three dimensional space, where every point in my three dimensional space is defined with the coordinate triplet $(x1, y1, z1)$. We have the same type of system today for the Global Positioning System, however, that is a two dimensional grid on the surface of the planet, using two dimensional coordinates $(x1, y1)$, to define a point precisely anywhere on the planet's surface, mostly helping our nautical navigation. In the next few pages I will investigate our three dimensional existence and more.

Why all these dimensions and why is studying them so important ? The answer is simple. This is the basic environment for humanity. All our activities takes place in our three dimensional environment. We can build cars, bridges, shopping malls, etc., because our knowledge and the facts of our 3D environment.

Next I will introduce here something that I will call a "thought experiment". I am not the first one who did this, so no credit is due me. Early on in history, about 2500 years ago, Plato (in the allegory of caves) and then Einstein, in his analysis of trains traveling at the speed of light, used a "thought experiment" because they both were investigating events that humans can't do in reality, but we (humans) can do it in our human mind.

Part2 - Chapter 11
Fundamentals, Matter vs. Mind, Space, Time, Energy, Directional Theory

Fundamentals, Matter vs. Mind, Space, Time, Energy,

Directional Theory (cont.)

Let us investigate my "thought experiment". Picture a two dimensional existence. A society of two dimensional people living on the surface of your dinning room table. How are you planning to communicate with them? (are you even planning to communicate with them at all?) Their movements are confined to two dimensions, and they can't see you or observe you in any way. They also can't leave the surface of the table, that is there universe.

Interestingly we can send a signal to them using light. We can have a flash light and turn it on and off, trying to signal those two dimensional existences. That is another whole new subject — how they will recognize, analyze, interpret and maybe try to respond to those signals. What about if we make some crop circles in their fields? One can ask what is the reason for this absurd example? Think about it, what if those unexplained UFO-s may be coming from the 4th or higher dimensions? Isn't it the same that we observe only lights and crop circles?

If it sounds like a speculative nonsense, I have to remind you that before we realized that the Sun is the center of our solar system and not the Earth as we thought, the theory came from the world of mathematics instead of observations. It looks like our mind can mirror some facts of the universe some times way ahead of our 3D physical capabilities. Humans were thinking about flying at the time during the Greek area (we find thoughts in their mythology), about 2500 years ago, but then they definitely didn't have the capabilities to fly from continent to continent as we can do today. Our mathematicians today can construct an N dimensional area, yet of course we don't have an observable reality to mirror their theories. (Sounds a lot like Copernicus!) So, why am I so hung up on N dimensions?

The answer is simple — because I think that is the proper way to look at our universe and try to find out where are we in this "big world". Before I will go and investigate further to see the possible reality of the N dimensional world I will revisit and extend my Chart 7. This chart will be the basic concept of the flexibility of the connection of Matter, Mind, Space, Time and the Directional Theory. I will go on then to investigate this new chart, then continue on with the McKaneist possible definition of "The Universe".

Part2 - Chapter 11
Fundamentals, Matter vs. Mind, Space, Time, Energy, Directional Theory

Fundamentals, Matter vs. Mind, Space, Time, Energy,

Directional Theory (cont.)

Here is the more complex Chart 22, the extended concept of Chart 7, the connection of Matter, Mind, Space, Time, Energy and the Directional Theory. I will analyze how the components in this chart relate to each other.

Chart 22

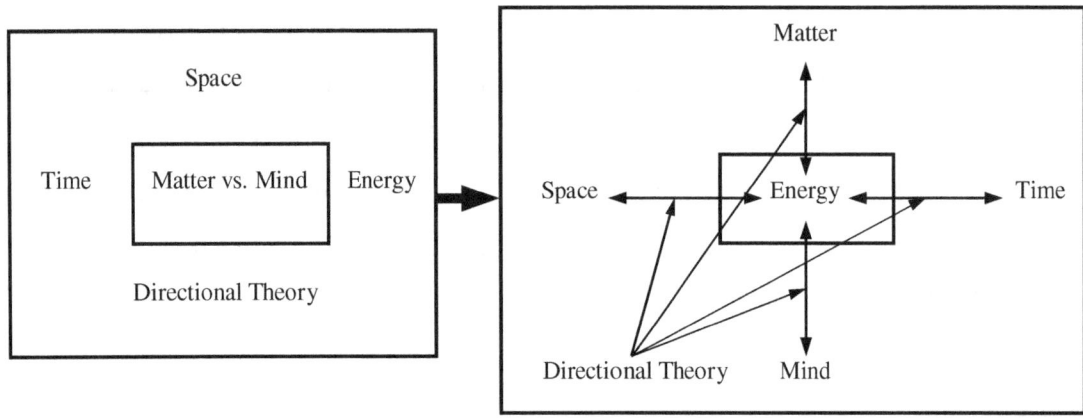

Please notice here that my Chart 22 is more complex but it is still in the two dimensional world. I will try to extend the dimensions later in my arguments.

This is my next **Bold** definition, as Chart 22 shows: **Matter, Mind, Space, Time, can transfer into each other through Energy based on the Directional Theory.** The level of energy and the sophistication of the mind determine the translation. This is a quantum leap compared to all existing views of today. It makes sense and it is logical for me as a philosophical observation, and probably will take many years for science to prove it or disprove it. The only example I have, one of the biggest secrets on Earth, is the "Philadelphia Experiment". I can't tell you what happened there because I didn't witness the experiment itself, and the extreme secrecy around it can't let anyone see any written documentation. I am always skeptical with eye witnesses because in our country (USA) many people are willing to appear on TV for their "15 minutes of fame" and the "big bucks" that are coming with it, and these people come up with stories, that they were there or they witnessed something.

Part2 - Chapter 11
Fundamentals, Matter vs. Mind, Space, Time, Energy, Directional Theory

Fundamentals, Matter vs. Mind, Space, Time, Energy, Directional Theory (cont.)

The secret remains secret for now, (remember from the introduction Part 2 where I said that the "Unknown remain unknown" in McKaneism!) however, according to eyewitnesses the energy created on one ship messed up matter, space, and time. Of course our government officials explanation of this is that this experiment never took place and any ship with that name never existed. (Isn't it easier keep all of us stupid, if we only hear about the Sunday football game scores?) For my Chart 22, I will be interested in finding some experiments that can put some spotlight on the correctness or impossibility of my theory.

Of course I should not be so excited, because "**Only the survival of humanity**" is at stake. The more stupid we are, the more sure that we "shake hands with T-rex in Heaven!" My explanation is (if my Chart 22 is correct), humanity is in an experiment, and like many times in human history, "opens the door into the future". At which point we run into something unexplainable, get real scared, closed the door, sealed the experiment, and hiding it from everybody — also hiding from it's reality. This is human nature until someone with the knowledge comes around and can explain what happened, what went wrong, and how we can make it better, applying some corrections. Then at that time we (humanity) will resume our investigation into the future. My explanation is simple. I have to say first that I've only seen documentaries and speculations of that experiment, so my explanation covers only philosophical, hypothetical thinking based on McKaneism.

In that experiment we generated **Energy** that opened the door for **Space** and **Time** to alter the **Matter** from our regular daily three dimensional life. In that time because we had no knowledge how to influence the experiment in the proper way with the proper knowledgeable **Mind** — we ended up with a major unexplained disaster. I will leave this subject there for now, until some day someone can find out the facts.

Let me spend a little more time with Chart 22. Looking at the right side of the **new** representation of our old Chart 7, one needs to notice a huge change and difference in that chart. The new structure shows how energy is the focus of all changes. Anything we check or view in our life has a well defined energy level.

Part2 - Chapter 11
Fundamentals, Matter vs. Mind, Space, Time, Energy, Directional Theory

Fundamentals, Matter vs. Mind, Space, Time, Energy,

Directional Theory (cont.)

The "matter" with the continuously moving atoms inside define the energy level of things. The change in that energy level can open the door for our unknown universe. Let me move on now, and extend my directional theory, and add Chart 23 for representation. (A picture is worth a thousand words!) I argued earlier that every existing thing (energy, matter, mind space, time) follow the rules of the **Directional Theory**.

The complete definition of the **Directional Theory** is: **The Matter and/or Mind, Time, and Space are based on the level of Energy and can move from simple to complex, or complex to simple, or any other direction, either slow or fast or any speed.**

Here is a chart to represent The Directional Theory:

Chart 23

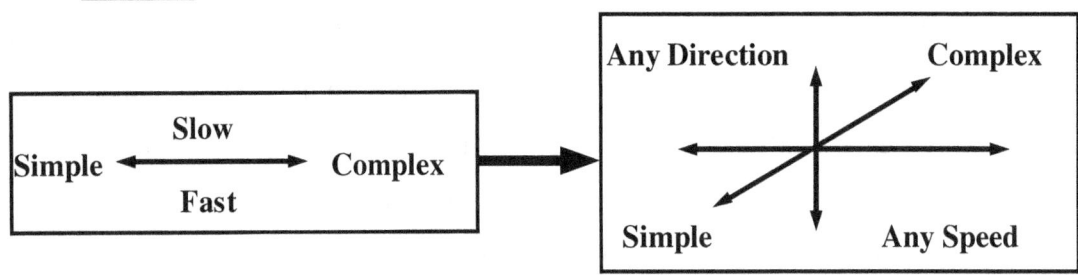

One can see the extension of the Directional Theory in Chart 23. I called it the **Directional Theory** because it is based on the observation of the physical world and the changes that take place in it. I was thinking of calling this "Theory" the "Directional Principals", but I realized that I have "hypothetical examples" that can be described as a theory but can't be called principles. It looks like every event in our world follows the "Directional Theory" but I can't exclude based on my observations and knowledge — I could have "hypothetical examples" that will make it only "Theoretical".

Part2 - Chapter 11
Fundamentals, Matter vs. Mind, Space, Time, Energy, Directional Theory

Fundamentals, Matter vs. Mind, Space, Time, Energy,
Directional Theory (cont.)

Here are few real examples: the development of a human being is the "**Simple to Complex, Slow**" directional process, the extinction of the Dinosaurs is a "**Complex to Simple, Fast**" process, the fusion process in our Sun is a "**Simple to Complex, Fast**" process, (Hydrogen to Helium, from a simple to a complex element.) for a "**Complex to Simple, Slow**", when a human being dies, we can observe the disappearing process of the body in the grave. (Philosophy and it's rules have no sad emotions as we do!)   One can observe a flooding river that moves out of it's banks as an **Any Direction, Any Speed** directional process. Now comes the extended part that kept these rules a "Theory". One can imagine the directional move from "one dimension to the other" with **Any Direction, Any Speed**. (For example moving from 3D to four dimensions.) Looks and sounds impossible (as of the year 2003), but in the logical thinking mind it stays "hypothetical, theoretical, but not impossible!".

Einstein constructed a "thought experiment" where he took the third coordinate z_1 and declared that z_1 would be 0 ($z_1 = 0$), Then he introduced the fourth coordinate called Time, t_1. So in this system, every point has coordinates (x_1, y_1, $z_1=0$, t_1) to describe that universe.

In this experiment he introduced a "four dimensional world" where the third coordinate is always zero. So "hypothetically or theoretically" I can see the logical possibilities of moving from one dimension to the other, of course in our current science nobody is focusing on this scientific or logical possibilities, (at least as far as I know at the time I am writing this book!) because it is science fiction and a non-science in most people's mind. One needs to understand that in our current emotional world, a scientist can be destroyed in his or her career if that scientist deviates from a main street of science.

This is the main reason that this world needs some "nobody" like myself, who does not care about that false value system. (More about value systems in Chapter 17!) If something is a logical possibility, than that something is a logical possibility, regardless of who says what.

Part2 - Chapter 11
Fundamentals, Matter vs. Mind, Space, Time, Energy, Directional Theory

Fundamentals, Matter vs. Mind, Space, Time, Energy, Directional Theory (cont.)

In my arguments I will go as far as an R dimensional Universe, (where R is the set of the Real numbers!) and I will describe a new model of that Universe in the next few pages. Imagine the impact of this new model on our current Matter vs. Mind arguments. On one side where "The Superior Mind", God created a static 3 dimensional universe, (Mind over Matter) or perhaps the other side where the 3 dimensional universe is evolving from the "Big Bang" (Matter over Mind). Both of these theories look extremely illogical, non-working, simple minded, impossible to prove, designed by "man" to mislead or dazzle another "man". Both reassemble the previous thought that humanity needs a "Birth to Death" (with Heaven) system to feel emotionally comfortable. How about feeling comfortable with reality for a change!

I have to continue with this thought process of describing the universe for a little longer here. Two popular thought processes exist today, that the "Universe" either was "Created by a Higher Power (God), who still is overlooking and controlling every aspect of it, or the other is that a dynamic physical event such as the "Big Bang", (described by the language of mathematics) started the existence of our current universe around 15 billion years ago.

The first one is based on the "good old, believe you me!" concept that we have seen many times before. The second one is based on a number of mathematical formulas, that 99.99% of the planet's population can't understand nor comprehend, so the scientists just tell us "believe you me!". One can see that both explanations are based on the "emotional belief" concept instead of logical thinking. Here are a few logical (common sense) questions for both!

Here are a few questions and logical arguments for the **first concept**:

Where is the "higher power" now? Where is Heaven? Why does "He!" not want to talk to us anymore? Why did "He" create this miserable 75 to 100 years existence, full with diseases and a falling apart human body? (I can come up with a better design concept, and I am just a nobody!) Why do we need to live miserable to get to heaven?

Part2 - Chapter 11
Fundamentals, Matter vs. Mind, Space, Time, Energy, Directional Theory

Fundamentals, Matter vs. Mind, Space, Time, Energy,

Directional Theory (cont.)

Does not these facts represent a cruel creator, having a field day at our (humanity) expenses and our suffering? A church representative (priest of any religion) will tell me that I am already heading to Hell for these questions. So, we can't ask a "higher power" any doubtful questions? There goes the freedom of speech, right out the window! Maybe I don't want to go to heaven then, where there is no freedom of thinking or no freedom of speech. Why the higher power doesn't answer my questions, like sit down with me for a cup of coffee and straighten out my confused mind?!

Why can't we find any trace of "Him" with our telescopes and radio telescopes looking 15 billion years into our universe? And finally: Why does the church representative (priest of any religion) answer my questions instead of "God"? Can't he speak for "himself"?, or I have to **"believe"** that God talks to me through the representative? (a priest of course, in a private session, in a telepathic way, so it can't be realistically observed, recorded, or video taped?) How convenient for me that they do all the work, and I just have to show up in the church on Sunday (with my wallet) to say a few prayers and my road to Heaven is a joyride!

Here are a few questions and logical arguments for the **second concept**:

What happened before the "Big Bang"? (They say all matters of the "Universe" compressed into one point and exploded!) Are we describing the "Universe" with the "Universe"? If we can observe 15 billion years into the past, what is outside of the 15 billion light years radius? (Maybe that is where God's world begins?) So, I am not interested in the 2 volumes of mathematical formula, simply give me an answer about what is beyond the 15 billion light years radius "basketball"? If you are a "real!" scientist you will tell me "I don't know!". How about another "Big Bang" for the next 15 billion light years radius "baseball"?

How about many other "Parallel Big Bangs" and "Parallel Universes"? Let me answer it for you, we (humanity) don't know! We are sitting on this small planet in the huge "ocean" of the Universe and try to be so smart.

Part2 - Chapter 11
Fundamentals, Matter vs. Mind, Space, Time, Energy, Directional Theory

Fundamentals, Matter vs. Mind, Space, Time, Energy,

Directional Theory (cont.)

Seems like an ant trying to figure out human rocket science! In summary one can see that both concepts, in the process of searching for answers, create a comfortable answer for our emotions.

We (humanity) just can't deal with our own Reality!, or Can WE? Highly unlikely that in this huge size of the Universe we can find the absolute power (God), that coordinate and control all the happenings. Most likely then that "man" invented some stories to control other "men's" minds. McKaneism is searching for the truth. In McKaneism I don't accept any goofy explanation of anything that we can't observe by our senses or with our "extended senses" or that we can't recreate in a "lab" experiment. Theories and logical conclusions are very tricky, (Big Bang?) so I will keep my eyes on them.

You will be surprised, maybe even angry from time to time, when we discover the misleading nature some of our previous philosophies, made to manipulate people and societies. I will address this issue in a more comprehensive way in Chapter 14 and 15, where one can see that all these theories coming from the control, organization, power distribution, and social respect (disrespect?) directions that one "man" wants to gain from another "man". ("man" here means "mankind" so no offense for "woman-kind") **McKaneism is for humanity (for all) without any discrimination to anyone !!!**

In the next few pages I will introduce my concept of the "Universe", with the precondition that it is only a logical, theoretical model. I am not saying (like the other kinds), that this is the description of the Universe, but instead I am building a logical observable model, that looks sensible and seems to describe some of our world. I will leave it on our scientists to take this model into their "lab" and analyze it, experiment with it, and prove me right or wrong! This model is intended to be the next concept for humanity. It definitely fits into the philosophy of McKaneism, however, that doesn't make it right or wrong! I mentioned before that I will introduce an "R" dimensional Universe. In my model I will use some mathematical language, but it won't be "the two volume, written to the other five mathematicians on the planet".

Part2 - Chapter 11
Fundamentals, Matter vs. Mind, Space, Time, Energy, Directional Theory

Fundamentals, Matter vs. Mind, Space, Time, Energy,

Directional Theory (cont.)

When one tries to represent multi-dimensions, one needs to use this "simple" mathematical language. When I say "simple", one can see, as I said many times in Part 2, it simply means that you please remember your four or more years of advanced college or university schooling. Sorry, with a 5th grade reading level, you are out! (Not my fault, you elected not to be smarter! The school and library doors are open for everybody in the USA!)

Let us move on then and view the first model that I will try to describe — my view of our "3D Universe", then I will travel a little further and open our mind for a new concept of the "R Dimensional Universe". I will try to make my introduction "simple" and logical as possible. I think this topic is very logical, however, not too "simple"! So let us see how this new model stands up to the investigative mind!

"Three Dimensional Universe" (or 3D for short), is a world we are living in every day. The concept is coming from the Greek time from the geometry of Euclid. Picture a point in your living room. For example, a "well defined right front corner of your TV stand"! Now assign 3 coordinates to that point, $(x0, y0, z0)$ where $x0=y0=z0=0$. We just created a starting point for our 3D universe. So the next three axis (Chart 24) begins the "Universe" from your living room, a well defined right front corner of your TV stand"! One can see with a small imagination that "every point", like any "point" on any piece of furniture in the living room, can be described with three other coordinates. Here is our chart:

Chart 24

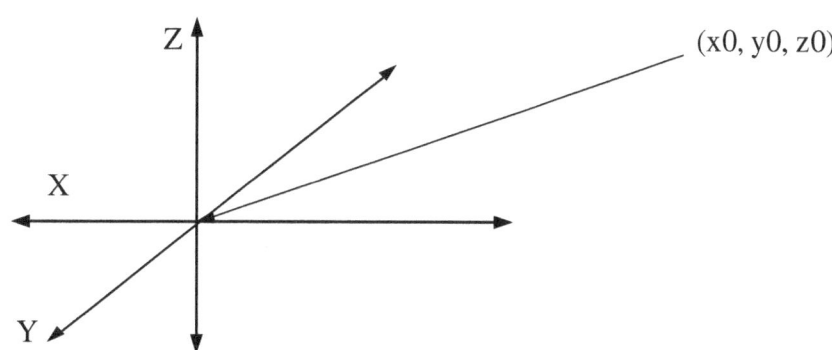

Part2 - Chapter 11
Fundamentals, Matter vs. Mind, Space, Time, Energy, Directional Theory

Fundamentals, Matter vs. Mind, Space, Time, Energy, Directional Theory (cont.)

So I can assign number triplets for every existing object, including atoms in the air in one's living room. With a little extended imagination, one can realize that if I extend these three axis, I can describe all "points" in my neighbors living room, then continue on to the corner gas station, then to the tallest building in my city, then to the capital of my state, then all the states in the USA, then all the points in planet Earth, and then all the points in the Solar system. I guess one gets my point in my reasoning. I am not painting this picture to be a smart one; next, this example has a very far reaching consequence.

If I place this extended 3 axis center in the center of the "Milky Way Galaxy", then I can describe every point in the galaxy. (I see that the "Math Guru"-s are starting to disagree with me, because a different "construction" other than Euclid's is out there, but stick around with me for a while!) Here I delineate a statement that with this model I made a "snap shot" of the "Milky Way Galaxy", defining every point in my model! (Here "defining" means that I know exactly where every point is compared to my center zero triplet.) So far not bad, however, I have some major problems here.

The first one is this: Where is the true center of the "Universe", so I can place my model in the center so I can define the "Whole Universe" compared to my zero point? My answer is very simple: I don't know where the center is, and I don't think humanity can answer that question either. The reason again that we try to look for a center is because that is the way of human life. First the planet Earth was center of the universe, then the Sun, then we find the "Big Bang" starting point, or "wherever God lives in Heaven". For example, for navigation the longitude=0 runs through London, a man made decision, because in our navigation history the English people always played a leading or important role on planet Earth.

For now I can make a strong argument, that: **For the next few millennium of humanity we can place the "Zero Point Triplet" in the center of the "Milky Way Galaxy", and we can live with this reality**, very easily without trying to "guess" the impossible, or unknown where is the center of the universe.

Part2 - Chapter 11
Fundamentals, Matter vs. Mind, Space, Time, Energy, Directional Theory

Fundamentals, Matter vs. Mind, Space, Time, Energy,
Directional Theory (cont.)

Our (humanity) knowledge is insufficient in the year 2003 to answer that question. If one tries to argue on my last point, one needs to understand that we have many hypothetical explanations for our universe, but because of our short time observation, it is hard to say which one can be correct. My 3D description of "Euclid's geometry" can be debated with the "parabolic geometry". We know that the galaxies are moving away from each other, (that is one observation based proof for the "Big Bang" theory), however, we have other theories that the galaxies are fluctuating, moving further away from each other then change course and start moving closer to each other. These moves can take millions or billions of years, so we have a way to go with our observations.

In this ever changing "Universe", that humanity seriously analyzed only maybe in the last 50 to 100 years, it is hard to conclude based on the short time observations, what can or could happen in the millions or billions of years.

One can start to understand and appreciate the flexibility of the philosophy of McKaneism, realizing that we can live in a relatively well established reality without finding the beginning or end of the universe. In later chapters I will argue that if we master our existence in the "Milky Way Galaxy", then we can start looking out and trying to discover the rest. For one to relate to the size of the problem: even traveling with the speed of light, it could take 100 thousand years to cross our own "Milky Way Galaxy". (It is about 100,000 light years in diameter, and 10,000 light years in height) According to our scientists, the next closest galaxy is about 2.5 million or billion light years away — traveling with the speed of light. So based on reality-based observations, it is impossible to conclude on the origin or behavior of the universe.

The second one is this: I established a model that covered our own galaxy as a snap shot. We all know snap shots, they are like a picture taken with a camera. I also described in Chart 22, that energy plays a big role in existence, so it is easy to see, that a snap shot can't work in an ever moving universe. No problem, it is time to extend my model to incorporate the constantly moving matter and mind.

131

Part2 - Chapter 11
Fundamentals, Matter vs. Mind, Space, Time, Energy, Directional Theory

Fundamentals, Matter vs. Mind, Space, Time, Energy,
Directional Theory (cont.)

We did see so far that out of all the "players" in Chart 22, we addressed Energy, Matter, Mind, Space. So it is easy to guess that Time is the next element that should enter the picture. In the next chart, Chart 25, I will extend my model with the introduction of Time. I will try to stay with the concept that I place the "Triple Zero" coordinates in the "center of our own galaxy".

Somehow I think it is easier to picture our "subset" of the universe if we think within the boundaries of our own "Milky Way Galaxy". We know from our observations and the science of our mathematics and physics that our Sun is about 26,000 light years away from the center of our galaxy and rotating around the center as our Earth rotates around the Sun. So there is the energy and movement that I will incorporate to the next chart. My new extended model will describe our existence, and after the introduction I will analyze some of the questions that I have no answer for. The answer has to come from physics, with the help of mathematics, but one needs to remember, that observation with our senses, and extended senses or an observed "lab" experiment, are the only proofs that I think we should consider in a realistic philosophy or any realistic thought process.

For the "Three Dimensional Universe" with Time, I introduce the $t0$ time coordinate. The original center $(x0, y0, z0)$ will change to $(t0, x0, y0, z0)$. Now, if I pick a point in my new "3D + Time Universe", like $(x1, y1, z1)$ with the change of the $t1, t2, t3$, time coordinate changes, I can describe the location of my triplicate $(x1, y1, z1)$ point in the ever changing "3D + Time Universe". The coordinates will look like (for this example), $(t1, x1, y1, z1)$, $(t2, x1, y1, z1)$, $(t3, x1, y1, z1)$.

Now my model is complete for our 3D Universe and Time, no surprise, is corresponding to Chart 22. The **Matter** and/or **Mind,** that is our "point" defined by the coordinates; **Time** and **Space** follow the **Directional Theory** (arrows **a** and **b**) based on the **Energy** level of the Matter and/or Mind.

Part2 - Chapter 11
Fundamentals, Matter vs. Mind, Space, Time, Energy, Directional Theory

Fundamentals, Matter vs. Mind, Space, Time, Energy,

Directional Theory (cont.)

This is the representation of the example in Chart 25.

Chart 25

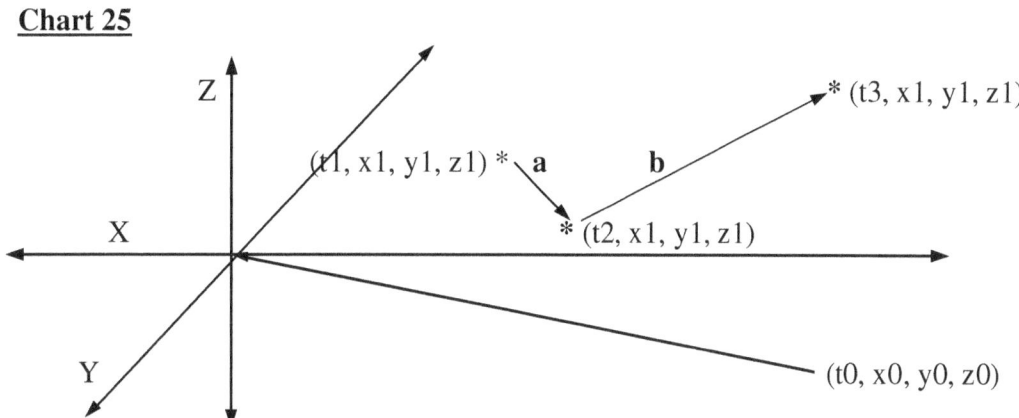

I will continue with a "thought experiment", then I will analyze some open questions about my model of the universe. The next "thought experiment" will introduce some interesting issues and connections through my new model. One can think for a moment that we have a very large supercomputer. Taking all the coordinates of the points in our galaxy, one can see that we can have all the knowledge of all existence and all events within the galaxy moving in a time line. Of course it is only hypothetical, because the number of points even trying to describe a simple small object like a table, it still can be very large. However this "thought experiment" raises the issue that hypothetically, is it possible that an intelligent mind can take control, or can gain an understanding of the whole galaxy around it, given the condition that the intelligent mind is capable of storing and modeling that large amount of data? Now, let me introduce some of the questions that this model raises in my mind.

I think taking the universe with the mechanical slow moving objects and events, my model describes the "universe" (remember we narrowed it down to the size of our own galaxy). However, when points start to move at the speed of light, I think we will run into some unanswered questions.

Part2 - Chapter 11
Fundamentals, Matter vs. Mind, Space, Time, Energy, Directional Theory

Fundamentals, Matter vs. Mind, Space, Time, Energy,

Directional Theory (cont.)

I also don't know how this model stands up for the different geometric systems. Seems to be working in Euclid's geometry, I don't know the rest. Regardless, I will extend this model to complete my hypothetical universe, that makes sense for me from the standpoint of philosophy, but I am sure that more investigation is required from other sciences.

Next I will re-introduce my structure of Chart 2 using the **new "3D + Time Universe" model** from the previous pages. In this new Chart 26 I will follow the same concept that I had in Chart 2, to take the Matter/Mind from a simple to more complex representation.

Chart 26 represents the McKaneist "3D + Time Universe" where all the matter/mind exists and all the events take place. (I still would like to stay on the "Milky Way Galaxy" level.) I will add a final logical extension to this model based on our previous hypothetical investigation into a potentially greater than "3D + Time Universe". We don't have to stop at 3D, however, we (humanity) don't have any observation or experience in more than 3D existence — more than that, even if I see the logical possibility, I can't present any explanation other than the hypothetical model of Chart 27 following Chart 26 in the next pages.

Before I leave this chapter, the last logical step to describe (the theoretical) "**The Universe**" (Chart 27) where our "**3D + Time Universe**" (Chart 26) is a subset of "**The Universe**". I can't prove this existence because it is probably beyond our 21st century knowledge, but this next extension makes a lot of logical sense based on our mathematical knowledge.

Part2 - Chapter 11
Fundamentals, Matter vs. Mind, Space, Time, Energy, Directional Theory

Fundamentals, Matter vs. Mind, Space, Time, Energy,

Directional Theory (cont.)

Here is our "World" in the "Three Dimensional (3D) + Time Universe" model:

Chart 26

t1	x1	y1	z1	In this chart "t" the time coordinate, "x, y, z" the three coordinates of the three dimensional universe. "R" represent the "Real Numbers" from mathematics, and 1,2,3, .. are the sequence numbers for the coordinates.
t2	x2	y2	z2	
t3	x3	y3	z3	
.	.	.	.	

"R" is the one to one relation between the linear line x and the possible numbers that can be represented on x from –infinity to +infinity.

tm	xm	ym	zm	**m for mechanics**
tp	xp	yp	zp	**p for physics**
tc	xc	yc	zc	**c for chemistry**
tb	xb	yb	zb	**b for biology**
tp1	xp1	yp1	zp1	**p1 for psychology**
ts	xs	ys	zs	**s for sociology**
ti	xi	yi	zi	**i for interactions of intelligent species** (as far as I venture to see in McKaneism !)
tR	xR	yR	zR	**R** for the "limit" of our "3D + Time Universe".

Part2 - Chapter 11
Fundamentals, Matter vs. Mind, Space, Time, Energy, Directional Theory

Fundamentals, Matter vs. Mind, Space, Time, Energy,

Directional Theory (cont.)

"The Universe" by McKaneism:

In this chart "t", "x, y, z", 1,2,3, "R", same as Chart 26, xx1, xx2, ... xxR the extended coordinates of the dimensions 4 to R.

Chart 27

t	x	y	z						
t1	x1	y1	z1		xx11	xx21	xx31	.....	xxR1
t2	x2	y2	z2		xx12	xx22	xx32	.....	xxR2
t3	x3	y3	z3		xx13	xx23	xx33	.....	xxR3
.	.	.	.						
tm	xm	ym	zm	m	.	.	.		.
.	.	.	.						
tp	xp	yp	zp	p	.	.	.		.
.	.	.	.						
tc	xc	yc	zc	c	.	.	.		.
.	.	.	.						
tb	xb	yb	zb	b	.	.	.		.
.	.	.	.						
tp1	xp1	yp1	zp1	p1	.	.	.		.
.	.	.	.						
ts	xs	ys	zs	s	.	.	.		.
.	.	.	.						
ti	xi	yi	zi	i	.	.	.		.
.	.	.	.						
.	.	.	.						
tR	xR	yR	zR	R	xx1R	xx2R	xx3R	.....	xxRR

I said before I am not a mathematician or physicist, but a philosopher. I will leave the closer investigation on those scientists, to prove or disprove the validity of my reasoning. Also I can't assign the complexity of the Matter/Mind to the right side of the chart because we (Humanity) have no experience or observation in any possible higher than 3D + Time Universe. These charts are representation of a philosophical concept. Just like I can draw a concept car — that doesn't mean that I also can build one. Let us move on and investigate Mechanics, Physics and Chemistry in the next chapter.

Part 2 - Chapter 12
Mechanics, Physics, Chemistry

Mechanics, Physics, Chemistry

My next subject will be the matter organized into different forms and the rules that apply to the matter within Mechanics, Physics, Chemistry, of course through the eyes of a philosopher. We did see in the previous chapter where the complex formation of the matter fits in our "3D + Time Universe". (I can't say much about the extended view of "The Universe" Chart 27 being reality based, because I think it is a new concept with no observations so far). I will stick to the "3D + Time Universe" for my reasoning, and also "shrink down" to the "Milky Way Galaxy" for an observable human reality, but the "general rules" may apply to any universe (for example "general rules", like the "rules of gravity"). Trying to stay realistic is a large enough order to investigate our galaxy as a "subset" of the "3D + Time Universe".

I think as far as **Mechanics** I can't add too much to my previous observations. Briefly Mechanics as a subset of Physics has one of the most useful applications in our everyday life. The design and building of the bridges, buildings, cars, ships, airplanes etc., (I don't think I can fit into this book all the items that are used in our daily life, that the rules of Mechanics help create, even if I have thousands of pages.) Karl Marx had some interesting logical references for the matter of mechanical movements, as a "most simple" out of the complexity of the matter. Interestingly, none of the religions have an argument with Mechanics, or for that matter, of the facts with Physics or Chemistry. Interestingly, those "Religious" thought processes aren't interested in the description of the universe until the level of Biology coming into the picture. Each of them has some kind of "Universe Creation" theory by one or more Gods, but no detailed explanation of Mechanics, Physics, or Chemistry. I guess "God" left that part to the scientists.

More interesting than that, in the 11th to 13th century Europe, practicing Chemistry was a science in hiding. The "Church" that was a leading power at the time didn't like "disturbing sciences" (like alchemy) to try to explain "God's World".

Physics is a different issue for me. It has been an all-time favorite of mine. Once upon a time I actually wanted to be a Physicist.

Part 2 - Chapter 12
Mechanics, Physics, Chemistry

Mechanics, Physics, Chemistry (cont.)

I did extensive study, reading in Physics, and picked this science as my elected subject for my High School final graduation exam. ("Looks like, it wasn't in the stars that I was to become a Physicist!?").

However, here I will introduce a few bold thoughts about some theories of our current Physics as a science. Some day some very smart scientist will look at my description of "**The Universe**" and discover the major strength of it, the **infinite possibilities** that is built into it. For me, it isn't a surprise, I didn't invent this, only through observation I tried to describe what is around us. The better my observation, the closer I get to the real world. Let me get back to Physics.

In our human life, we experience an "interval" type of living within our lifespan. Following this observation, we (humanity) tend to set an interval for everything around us. (Later in Chapter 17 of this book, I will address the rule of logic: "Specific", "Interval", "General"!) We look around and see that humans live a lifespan between 75 and 100 years by our measurement of time. (I use average and have no intention here to argue about all those exceptions when someone lived a shorter or longer life, whatever reason!) One can see that in a human lifetime, we see "Interval" existence. More than that we see "Beginning", "Interval Existence", and "Ending". Interestingly these emotionally very strong observations reflect strongly some of our explanations of our surroundings, and our imagined universe ("Big Bang", "Heaven", "Hell").

My "3D + Time Universe" model isn't imaginary, it is reflecting my observation of our reality! In the subjects of physics we use mathematics, on the side of our observations, and the "lab" experiments to explain the selected topics. I just want to remind my readers of my previous statement that in the science of physics we have a "Big Bang Theory", and in religions we have some variation of the "Creation Theory". Please, notice here that we (humanity) need a starting point of the universe as we have a starting point in our own life! "Big Bang" or "Creation" also address an ending of the universe. (Examples: Big Bang: the mass falls back to one point and explodes again, Creation: Anti Christ, and end of the world.)

Part 2 - Chapter 12
Mechanics, Physics, Chemistry

Mechanics, Physics, Chemistry (cont.)

As I argued earlier a comfortable "Birth to Death" Universe mirroring the human existence. All explained well, no surprises.

The concepts of McKaneism will surprise you, (if it didn't do it yet!) how about no start and end to "The Universe" as we know it in our human existence.

If you look into my formula of the universe, Chart 27, one can realize that "The Universe" can constantly change from one form to the other; it doesn't have to start and end, and only the level of existing energy governs the events that take place. What a bummer — the universe, it is not necessary to die with us. Currently we have no observable or "lab" experiment proof that "The Universe" has any starting point as suggested by the "Big Bang or Creation". Nobody was there, from our human race, to observe it, make a video or pictures of it, or recreate it in a "lab" environment, so all we do is just speculating.

I already had my argument with "Big Bang", how about I say that simply a "4D + Time Universe" transformed into a "3D Universe" — go ahead and prove me wrong with videos or observations. In the standing of reality, it is impossible for the 21st century human being!

The same holds true for the creation theory of the Bible. Who was there to document (video, pictures, film – not a conjecture document) the event that took place? The answer is nobody! Mostly we are operating on the "Believe you me!" premise.

My answer is: **Not an Option !!! The rule of reality is that we (humanity) have to observe with our senses or extended senses (telescopes, microscopes, cameras, etc.) or recreate the events in our "laboratory" environment to have valid, realistic conclusions.**

My next interesting observation in Physics, is that we "must" find the smallest building block of the matter. In our world of reality, we build our home from wood pieces, red brick blocks, etc. That is the way of things. We take components and build something together from those components. One more step and here we go: "The Universe" has to have it's own building blocks. We take the smallest pieces and put them together and we get bigger and bigger until we get something the size of the Universe!

Part 2 - Chapter 12
Mechanics, Physics, Chemistry

Mechanics, Physics, Chemistry (cont.)

Yet again, a comfortable man made concept that "The Universe" is built from smaller building blocks, just like our house. I have a different argument about that. I think that in our "magical world of the Cyclotrons", where we accelerate the particles to bump into each other, we create different conditions every time, even in a very closely controlled "lab" environment.

The reason for this is that when we are playing with our technology on the new discovery edge each and every time we fire up our Cyclotrons. When one gets to the level of nanoseconds or less, the experiment pushes to the edge of our possible observational limits. I am convinced by my own model of "The Universe" and all the reading I did about nuclear experiments in Cyclotrons, that we don't have the smallest building block in reality.

Only the level of energy determines the "appearance" of the created particles (matter) after the collision of protons. It makes sense based on the Directional Theory, that based on the energy level a different form of Matter appears. As far as my latest readings, each and every time we do this "lab" experiment we discover different small so called "subatomic particles". My conclusion is that we can experiment forever and we will find different particles each time, that the experiment changes just even a little bit. I can't prove my reasoning here, but as a philosopher, I think it is more sensible and logical that the matter in physics can exchange forms based on the energy level of that matter, than an existing "smallest to biggest theory". Anyway, a "smallest to biggest" is a very relative term, and lead us back to our human experiences, where we need to compare things to create our admiring emotions. (The tallest building on Earth, etc.)

I am convinced that when we apply a minor difference to each experiment, (probably not even measurable most of the time with our instruments; remember those experiments are at the leading edge, the best of the best of our science, and sometimes even the instrument for observation is experimental.) so we come up with different results. Then we conclude, that this is the building block, the smallest. (Of course we proudly giving fancy names for all those particles — "neutrino", "quarks" — sounds special, ain't it? Just can't stop running our world by emotions, even in the scientific field!)

Part 2 - Chapter 12
Mechanics, Physics, Chemistry

Mechanics, Physics, Chemistry (cont.)

I am very sure that we don't have the smallest or the biggest or the starting and the ending state of the matter. Much more logical that the energy level of the matter determines the state and form of the matter. We are dealing with such a high energy level in these Cyclotron experiments and such a small amount of matter, (like a few protons) that we can open doors in "The Universe" that we never even imagined before.

Each time humanity opens those doors the observations can be hard to explain. This is my "thought process" contribution to Physics from the view point of Philosophy, and I am sure as time goes by someone will prove that the observations and conclusions of McKaneism helped influence physics to explain and direct some of these experiments. I want to remind you from Chart 1 of this book, that Philosophy influences the sciences, with organizing their discoveries, and this organization can lead sciences to accelerate further and influence philosophy to reach the next level, prove or disapprove statements in both directions, and the cycle goes on and on.

The next science in this chapter is **Chemistry**. This science has a great deal of importance in the human existence. In the next chapter we will see biology where we analyze the vital signs of the living existences, however, Chemistry is the base for most of that. The on going chemical interactions in the living entities determine many of the actions, reactions, and behavioral patterns for them. Especially important is the "Carbon" based chemistry, a matter somewhere in the middle of the periodic table that has the great capability to connect to many others forming huge complex molecules as building blocks for the existence of living entities. Chemistry, as my "3D + Time Universe" Chart 26 shows, includes as a subset physics and mechanics — within physics as a subset of the subset. Without the mechanical movement of molecules they couldn't "bump" into each other and most of the time start a chemical reaction. Without the physical rules of the matter, the newly formed chemical compounds can't "sort out" the new structure of their nucleus and circling electrons balance. Not too many new surprises as far as these sciences relate to each other.

Part 2 - Chapter 12
Mechanics, Physics, Chemistry

Mechanics, Physics, Chemistry (cont.)

One more very noticeable power of chemistry. Scientists are discovering that in the human mate selection process that the Chemistry of the two individuals may outweigh (in fact) the box of chocolates, the roses, and sometimes even the diamonds. We investigated our senses and their functions in Chapter 4. According to the discoveries of the science of bio-chemistry, the physical attributes of a human being such as body shape, voice, body movement, etc., through our senses, signals our brain. Those brain signals are actually chemical compounds that influence our brain for the decision making of mate selection.

Somehow we are chemically programmed to respond for the other gender attributes. There goes the emotional gift of twelve red roses!

More than that, one can arrive at a logical conclusion, that our emotions are basically the reactions for our underlying chemical actions. So, it seems like the words Love, Happiness, Sadness, Joy, Anger, Depression, etc. are nothing else than a bunch of chemicals flowing through our brain. That is raising an interesting thought, that all our emotional words are different chemicals, or the same chemical with more or less volume, influencing our brain. The "good feelings" chemical are the same or are they different for each good feeling, like Calm, Comfortable, Happy, for good feelings or Restless, Miserable, Unhappy for the bad feelings. Interesting to discover that our chemical balances or out of balances can determine the good or bad functioning of the individual. We can see the importance of the chemical influences, on the human individual behavior in individual or social situations. We may not be as "Free" after all as we were told to believe. Our scientists are starting to discover that the food that we eat can affect our genetic make up.

Maybe all the chemicals around us, like pollution in the air, water, and food, can change our chemistry for the better or worse. I hope with my new system I contributed to the world of chemistry, and some of the scientists in the field of chemistry will be able to see his science in the new spotlight. I said many times that philosophy doesn't try to replace any of these sciences, it simply tries to influence them by organizing the existing knowledge of the 21st century.

Part 2 - Chapter 13

Biology, Psychology

Biology

In my "3D + Time Universe" biology and it's rules don't change too much compared to the observations that I introduced in Part 1 investigating biology. Let me start here with our new chart (Chart 28) to describe our new "3D + Time Universe" of Biology:

Chart 28

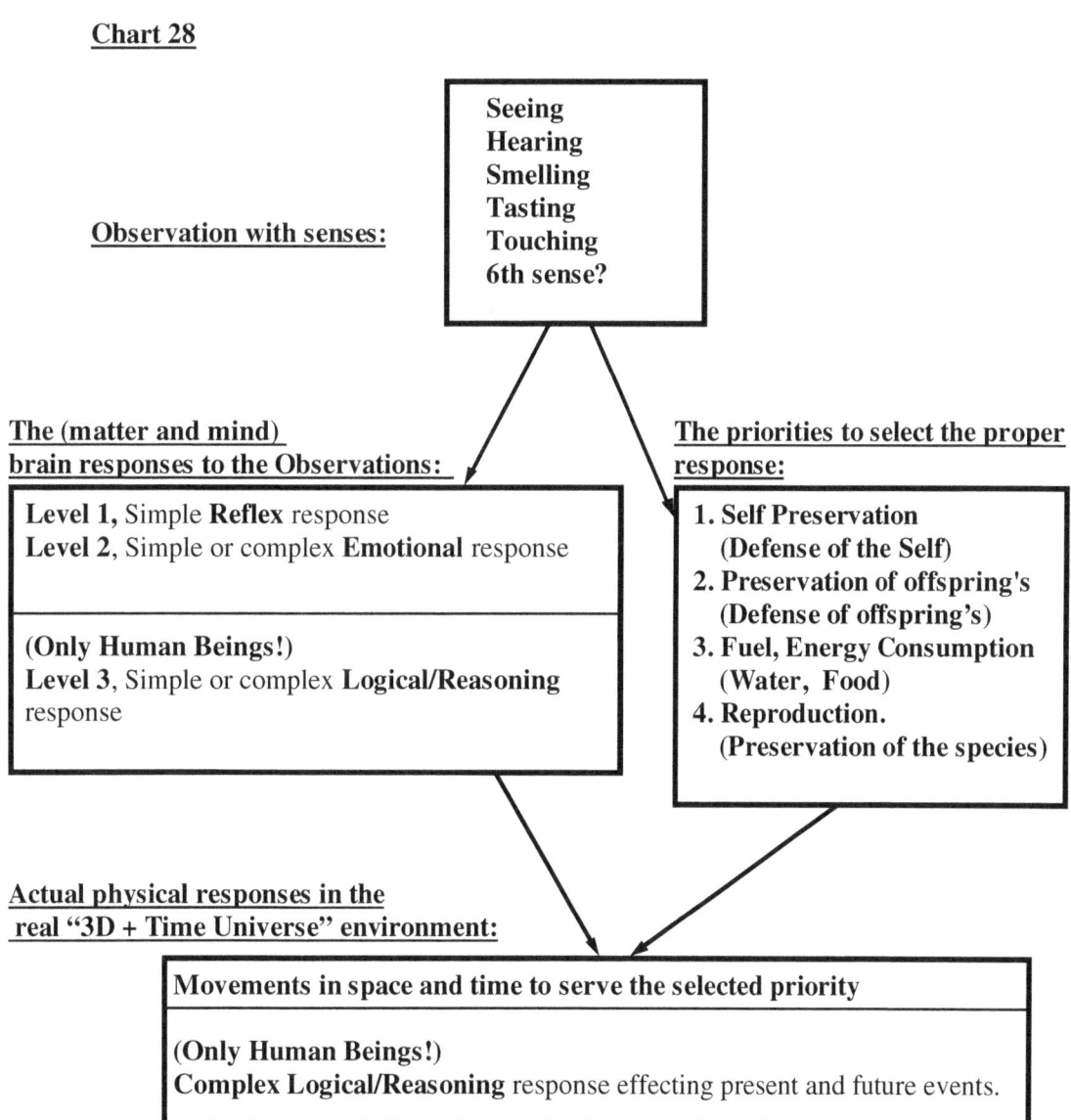

143

Part 2 - Chapter 13
Biology, Psychology

Biology (cont.)

Studying the living organisms and their attributes, that is what biology is all about. I will focus on "The Attribute of Living Entities Theory", because that has a very serious effect on our life, and the way we view our "living world". I also organized the groups of drivers and attributes (Chart 28), that I first defined in Part 1. It will help understanding their relations, and how they define the responses and reactions that living entities (plants, animals, humans) show in their environment.

We can see from Chart 28 on the previous page that only humans are capable of truly responding to the effects of the environment, not just altering the present, but altering the future as well. This is a very important attribute of humanity, to control the outside world. It brings to us a great responsibility for our actions. For example, in the early American West we (humanity) hunted the wolves to extinction because they were in the way of human cattle ranching. That is the power that humanity has within the Level 3 Logical / Reasoning mind, even if the decision is not always the correct one.

This can't happen in the animal world, that one species can make another extinct. (Only the change in the environment and the lack of the capability to adopt to it, can!) The fairness of the natural distribution system creates a well balanced environment of the proper food chain, with the important functionality of all players, with no discrimination at all. They are all equipped with the strength of survival skills, and the need and place in the food chain. (Anyone who tries to argue with this must realize that we humans are constantly pushing and squeezing their environment, where we take away their possibilities to survive!) What we need to understand is that the Biological Attributes define the possibilities of each living entity.

For example, birds can fly, humans can't. Fish can live under water, humans can't. Humans can artificially change the environment (even bringing extinction to every other living thing), plants and animals can't. One huge discovery here, that we will revisit in the level of sociology, is how these human attributes can be used by one group of people to rule the rest!

Part 2 - Chapter 13
Biology, Psychology

Biology (cont.)

The next important observation that I would like to make is that all the Biological Entities are built on the very complex structures of carbon based chemicals.

Why is it so important?

It is important because those chemical reactions, as a subset of their biological existence, determine many of the biological needs of an entity. Many of the functions of the brain that manifest as fear, or happiness, or joy, or many others, are very different emotional stages, especially for humans — these happen in the Level 2 emotional response level. The chemical reactions triggered by the sensations through our senses define the responses of our body (matter) and brain (mind) in the complex physical environment. Unfortunately, humans can go as far as using chemicals (called "Drugs", and I am not talking medicine here) to artificially alter their mind and create desired positive emotional states of mind, even if in the process they are slowly or sometimes quickly destroying the body and mind itself.

We need to understand that the "natural system", which means to create a well balanced planet Earth, also makes every living existence that can't or didn't develop the **Level 3**, **Logical/Reasoning** response (Chart 11) capabilities "**Earthbound!**". **They can't step outside their environment without losing their existence!** That happened to the dinosaurs, at the time of the meteorite impact on Earth about 65 million years ago.

Anybody can argue with the scientists about the "Dinosaurs extinction theory" but also anybody can observe the facts that we found the huge bones, and not one living species to correspond to those bones, so they are all gone — no matter how it happened!

This seems to me a fact that describes and finalizes my observations and arguments, that the Species without the Level 3 Logical / Reasoning mind doesn't have a chance to survive a fast changing environment. This observation and argument will be revisited and will be very important in a later chapter when I will make some arguments of the future of Humanity.

A few more interesting observations about the 6th sense of the living entities.

Part 2 - Chapter 13
Biology, Psychology

Biology (cont.)

The old Chinese medical chart describes an "energy field" around our human body, and they also use the power of "Chi" for healing the sick ones. In their fighting art forms they introduced their mysterious "Death Touch".

That is (according to the legends) a punch delivered at a special point to the body that can turn the positive chi into negative chi and the person dies. In this legend only the few special ones can cure the injured "body and mind", reversing the flow of chi to avoid fatal death.

We also know that some sensitive psychics described glowing lights or lack of them, as they described an "aura" around the body.

One more interesting short film I observed. This film had a small group of visitors of an African Safari stopping by and photographing a resting group of (12-14) lions, half way asleep, as lions spend about 22 hours out of 24. It happened to be that one of the 4x4 vehicles got stuck in the muddy grass underneath. As they tried to free the vehicle, the lions became very interested and started to approach the visitors.

The guide advised them not to step down from the vehicle. Was it a commotion that disturbed the lions, or maybe the change in the humans and the lions started to sense their panic? Maybe the lions sensed with their 6th sense, that a troubled, easy food item was around? Could it be that the "fear" triggered the change in the humans "energy fields" that the lions picked up? I don't have the answer, but I will offer the following theory for all these examples.

We know from our science of physics that one nature of matter has electric, magnetic and gravitational fields around it. If one takes a look at our chart it easy to see that physics is a subset of biology, so any rules in the physical world will be a subset of the biological world also. Following this logic, it won't be a surprise for me if in the future someone will be able to register any or all of those fields around us and also will be able to explain those affects on human biology.

A final note before we move on to investigate Psychology, that is if you are in the field of Bio-Chemistry and you think I left you out in my philosophy, I didn't!

Part 2 - Chapter 13
Biology, Psychology

Biology (cont.), Psychology

Your area of study perfectly fits into McKaneism between chemistry and biology, looking at our "3D + Time Universe" chart. The same stands for Bio-Physics or any other science. Remember McKaneism is intended to describe "The Universe" around us with everything and everybody in it, nobody or no knowledge nor thought process gets left out.

Psychology

Here we venture into the world of the human mind. Probably one of the most mysterious and most exciting areas of study in human history. I will address a few areas that I summed together under psychology. This investigation of mine will run around the "Self-realizing Mind". In mind studies we can find traditional areas as well as newly defined areas by my philosophy. I have no intention again to pretend that I can have knowledge of all these areas, but again, I don't need it either. I will address all these areas from the viewpoint of philosophy, as I did with all the other areas in the previous chapters.

<u>**Those areas that I am investigating are:**</u>
1. **Medical science, psychology, psychiatry**
2. **The extended "3D + Time Universe" Knowledge Chart diagram**
3. **The extended Attributes of Human Beings**
4. **Questions about our thinking or thought processes**
5. **My philosophical theories of the directions of the self-realizing mind**

<u>**1. Medical science, psychology, psychiatry**</u>

Humankind by the historical process separated Medical Science (the world of the doctors), Psychology, and Psychiatry. (By the way, they are all doctors, but because this topic is so big it takes more than one mind to absorb this huge amount of knowledge.) This profession requires one of the longest study. Some doctors are so good that they can cover more than one out of these three areas.

Part 2 - Chapter 13
Biology, Psychology

Psychology (cont.)

1. Medical science, psychology, psychiatry (cont.)

Medical Science is the study of the human brain as a physical entity, (by medical doctors) that goes back many thousands of years. Based on the belief system of the given society, they performed many (mostly brutal) cures on mentally sick individuals, that many times ended with the result of death of the ill person. Our modern medicine began in the 19th century, where instruments were able to measure the brain activities through electrical impulses and "map" our brain with special x-ray instruments. That part of mind study is realistic medical science and it keeps improving with ever improving technology. (using our extended senses)

Psychology as a serious modern science started in the Western world with Sigmund Fraud, an Austrian doctor, who discovered that our mind is focused on some thoughts that can have positive or negative effects on our health. He spent a great deal of time analyzing people's (his patient's) dreams and tried to understand the underlying problems that are invoked by those dreams. Even up to today he is considered the father of psychology by many. This is a very controversial area of mind study, that tries to use the power of the mind to cure the illnesses of the mind. One can find the same concept in the Oriental world in the far East.

I investigated the "Chi" (for the Chinese, "KI' for the Japanese), in Biology, the mystical power of the human mind that can be used to control, cure, help, or destroy our human body and/or mind. Interesting to notice that most martial arts of the orient searching for the "Harmony of Body, Spirit, and Mind", — like one of my favorites, "Aikido". (That thought of the Orient looks like it is matching my observation of the "Harmony of the Level 1,2,3 system" in our human existence (Chart 11)!) The difference is that they believed that one can achieve harmony through the control over one's mind in any environment.

I am strongly convinced that the individual needs a "proper social setting" (more in Chapter 14, 15) to achieve that harmony. If one investigates and analyzes some of the "Great Teachers" and their art form, one can realize that all the great achievers removed themselves from the society.

Part 2 - Chapter 13
Biology, Psychology

Psychology (cont.)

1. Medical science, psychology, psychiatry (cont.)

They were living in "chosen exile" from the society (up in their high mountains) practicing their martial art and meditating to strengthen and calm their mind. (No heavy traffic, long lines in banks or food stores or gas stations, and many other thousands of disturbances that take ones mind out of harmony in our modern western life!) I came to my own conclusion with observing these facts that we need an artificially designed Level 3-based society with Level 1,2 harmony on top of it, to achieve an environment where we all can follow the great and very enviable "Body, Spirit, Mind harmony" of those great (**few**) masters who introduced to us as a great "would be way of life", (**for all**).

Psychiatry, I think as a modern science, seriously begins with the Russian scientist Pavlov, who performed his experiments on dogs, and strongly believed that the conditioning of the mind through establishing chemical reactions he could alter the behavior of the mind. This is one of the fastest growing sciences of the 20th and 21st centuries as humanity realized it, that the human brain is a chemical, electrical instrument and with different medicines that we can alter the chemistry of our brain and cure the many illnesses of our mind. We discovered (our scientists) that many of the physically noticeable illnesses like head aches and many hard to discover hidden illnesses like having a "bad mood" or "mild depression", are coming from the chemistry of our brain.

No surprise for me, if one remember my Chart 2 or Chart 26, one realizes that Chemistry is a subset of Physiology, and of course this is within Biology. That can explain how chemistry can effect our (humanity) psychological make up. Psychiatry is a very serious medical science (all three of them!) and I think will be the one that will "Bridge" the human understanding between "Medical Science" — the physical world of the brain science and "Psychology" — the currently elusive science of the study of the many behaviors of the mind. All three of these sciences exist and go hand to hand in our 21st century, and because of the size of these knowledge areas, in our current century most of the time, requires an independent specialist to cover each.

Part 2 - Chapter 13
Biology, Psychology

Psychology (cont.)

1. Medical science, psychology, psychiatry (cont.)

We have doctors who are Brain surgeons practicing the **Medical Science** of surgery and doctors practicing **Psychology** or doctors practicing **Psychiatry**.

2. The extended "3D + Time Universe" Knowledge Chart diagram

The investigation of the "Human Knowledge" is an important part of McKaneism. Most philosophies, (but not all) try to answer some questions about the working human mind. Most of them has some sort of Knowledge Theory investigating how the human mind is learning and thinking. I will investigate the human thinking in Chapter 17. For now I will continue to extend on my Chart 14, Knowledge Chart diagram from Chapter 5.

This time I will construct a real life reflecting realistic "3D + Time Universe" Knowledge Chart diagram. This will be a natural extension of Chart 14 in the 3D world incorporating the concept of both Chart 14 and Chart 26.

One will see that this chart is a very complex chart, so I will try to represent the concept with a picture and descriptions — I will introduce the conclusions of my observations and studies around our human knowledge. I will pick some of my own skills to introduce the basic chart, then I will describe the nature and rules of the complete chart.

Just like every event that we saw before, the "Human Knowledge Chart diagram" also follows the Directional Theory. That means that our knowledge can grow or shrink in the time line. Most of us use about 5% (on average) of our brain capacity. The function to forget is "naturally programmed" in our brain as a defensive mechanism. We can gain knowledge every day through our constant, alert observations and thinking, and in the same time we can lose some of our knowledge, that we forget. Mostly, the rarely used skills that we learned are fading away in our memory. Aging also plays a great factor in the decreasing knowledge of the individual.

The "Biology" of the healthy human brain is that we have about 14 billion brain cells. (Fully grown healthy adult brain) Every day we are losing about 100,000 brain cells in the normal daily life.

Part 2 - Chapter 13
Biology, Psychology

Psychology (cont.)

2. The extended "3D + Time Universe" Knowledge Chart diagram (cont.)

If one has a high stress level or alcohol or any other addictions this number can double or triple or can be even higher. The fact is that our brain cells are the only one of those types of cells in our body that won't get replaced with new ones. (Some scientists think this can be a limitation of our highest age, because around 125 we lose so many brain cells, that our brain can't have a normal function anymore) These are the cells that connect in our brain, creating the foundation for storing, learning, and forgetting our knowledge. One will see on my Knowledge chart that we can't have negative knowledge.

That doesn't make sense logically. (Our mind is never empty!) One can have zero knowledge in some areas, but to have "total zero knowledge", it is impossible (except when one is dead!). One needs to remember that from the first second of any human life the baby already has the Level 1, native born (reflex) knowledge of where to find the food.

It is biologically impossible to separate the Level 1,2,3 Knowledge of the human brain. The Level 1,2 mostly chemical, emotional reactions. Level 2 being responsible for most of our emotional feelings. The stored knowledge is in our Level 3 (Gray upper area of our brain). The reason I made my argument that Level 1,2,3 is hard to separate, for example, one can think of one's baby and the "thought itself in the Level 3" brain can create a "Warm, Loving Feeling in the Level 2" brain. The functions of the three levels mix sometimes, however, the Logical / Reasoning thinking takes place most of the time in the Level 3 area of the brain.

Now it is time to construct the "3D + Time Universe" (Chart 26) with the Knowledge Chart Diagram (Chart 14), creating a new Chart 29. I will use part of my own knowledge diagram, choosing my three skills, as computer science, philosophy, and swimming. Let us investigate the following chart (Chart 29) and see:

How to read this chart?

Part 2 - Chapter 13
Biology, Psychology

Psychology (cont.)

2. The extended "3D + Time Universe" Knowledge Chart diagram (cont.)

Here is my (partial) extended "Knowledge Chart Diagram" in the "3D + Time Universe":

Chart 29

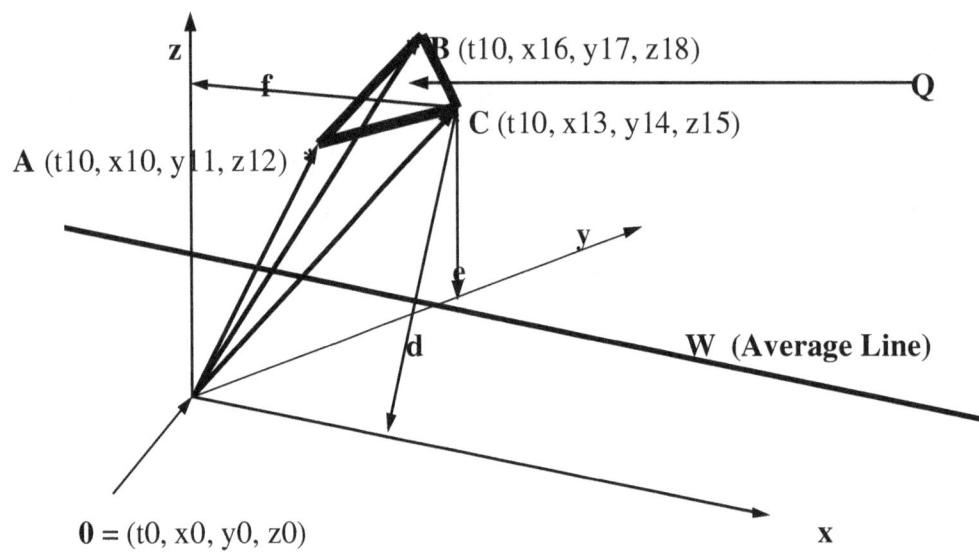

One can see that the concept is to assign a "**Point of Knowledge to Level A, B, C**" to the skills (**All** in a complete chart, **A to Z** as we say it in daily English or **k1 to kR** by the language of mathematics, where **k** is for the knowledge component **1** to **R** is where **R** is the well known **Real** numbers from the basics!)**,** that one individual may have. If we look at point **A, B, C**, they represent three (3) of my skills that I learned through my lifetime.

A = Computer Science, **B**=Philosophy, **C** =Swimming. For the example here, I draw a parallel to the **x** coordinates a **W (Average Line)**. (We have an existing **Average Line** to **y** and **z** also, but I don't draw those to avoid a more confusing chart!) It is easy to see that my **A, B, C "Point of Knowledge Level"** coordinates, all three are above the **W (Average Line)**. (Of course the same would be true for the "parallel to **y** and **z** **Average Line**").

Part 2 - Chapter 13
Biology, Psychology

Psychology (cont.)

2. The extended "3D + Time Universe" Knowledge Chart diagram (cont.)

It isn't any great achievement considering that I spent more than 25 years in **A =** Computer Science, more than 30 years with part time **B =** Philosophy and about 6 years as **C** = Swimming (as racing in the swimming team). So the above average level really not a special achievement. Probably anyone with the same years of practice would achieve the same level, or maybe even higher. However, the importance is the representations of the **"A, B, C Point of Knowledge Level"** skills.

The **"0 starting point and A, B, C, vector ending "** is my **"3D + Time (= 10) Knowledge Chart Diagram"** of my three skills that I chose in my arguments. One can also see out of the three skills, none are on the same level. (The vectors **A, B, C,** have different sizes).

The **C** = Swimming knowledge level represented with the **d** = x13, **e**=y14, **f**=z15, vectors, gives one the visual representation of the possible differences in the coordinates **x, y, z.** One can also see that if I represent **All** my skills, (basketball, tennis, geography, archery, etc.) I would end up with my **"Complete 3D + Time Knowledge Chart Diagram"**, (in one point of time) that is actually a "3D Object" in the "3D + Time Universe". It isn't a surprise, if one remembers that our mind is mirroring (self-realizing) reality or reflecting the real knowledge as a mirror.

One also can notice that the **Time** coordinate is **t10** in this example diagram. That represents the "relativity" of the knowledge diagram. **Time t10** is a well defined static point in my personal development, (snap shot) of my knowledge.

If we create the chart in the future, like 2 weeks from now (**t13**), it can represent the enhancement of the existing skills, or even a new arrow to represent a learned new skill. (One will see the relativity argument in Chapter 17). For now, an example looking at the same Chart 29: I may have picked up a little from another programming language in two weeks, so my A = Computer Science skill gets bigger (vector longer) or I learned how to play chess, which I didn't know before at all. So in the next "snap shot" in **t13,** in my **Time Line** I can have a more extended **"3D + Time Knowledge Chart Diagram"**.

Part 2 - Chapter 13
Biology, Psychology

Psychology (cont.)

2. The extended "3D + Time Universe" Knowledge Chart diagram (cont.)

(Extended with "one new arrow" **D** = Chess, and the "longer arrow" for A = Computer Science).

One can observe that with the "**Linear Time Line**" the **relativity** of the level of my knowledge comes into the picture. One can see the reverse example also if someone hit me on the head and damaged my brain, I can lose some of my skills due to the injury, so my "**3D + Time Knowledge Chart diagram**" will reflect a smaller "**3D Object**".

Why are all these investigations so important?

Many important implications come from this chart. The first one is that the "**Triangle Q**" is the "**connections between knowledge areas and comprehensions of logical thinking**" in our mind. (Mind mostly defined in philosophies as a working brain!) For example, I came to the self realization based on my chart (Chart 29) that I am better in individual sports than team sports. One can see that this statement comes out strange, when looking at Chart 29, and one can ask "how is that conclusion driven from that chart?"

Here is the "logical stream" of thoughts. I feel great when I am swimming individually, many laps alone in the swimming pool. In computer science, I performed at "Peek Performance" when I had an individual assignment. (Someone defined the task, then left me alone to accomplish it, of course, with a set timeline for ending.) Learning Philosophy I was always fascinated with societies that gave individuals a freedom within a society. (You guessed right, I don't like communism!)

So, in my "**Triangle Q**" every sign is pointing to an individual (in the individual vs. social) direction. One can see the straight forward conclusion as myself, that it is a more individual than team (social) player (in general) in my whole existence.

I will leave this thought process for further examination for the scientists of these areas, and many of us will be surprised as to how much valuable information they can get out of this new concept, that can affect our "individual development" and our "education system" as a whole in the society.

Part 2 - Chapter 13
Biology, Psychology

Psychology (cont.)

2. The extended "3D + Time Universe" Knowledge Chart diagram (cont.)

The second implication is for our education system. In our human societies we subject our young ones to many years of studying. (Usually from age 6 to 18, and another 4 to 6 or 7 years in specialized education /universities, colleges). One needs to understand that studying is hard work. Some of the subjects that someone likes can be fun to master, but most other subjects are hard work to gain the required knowledge to become ready for certain professions. We need to be careful that in the process of making studying fun, we don't ease out on the subjects that must be mastered.

One more interesting observation that I have to make here. I was studying Latin in my 9 to 11 grade high school. The reason was is that I hadn't decided yet to enroll in a medical field, but Latin is very important to know there. My Latin instructor was an excellent teacher and a great educator. One day he made the statement that "one can tell when one speaks to a person that **a person** studied Latin or not". It took me many years to figure it out what he trying to say with his statement. The fact is that when one is studying Latin, one studies the language of the famous speakers (Cicero, etc.) from the Roman era. The flowery language and the long complicated sentences that they used definitely impacts one's capabilities to think and comprehend — following and understanding long complicated sentences.

That is the secret that I understood many years later, that my teacher tried to tell me. This observation and discovery directly relates to the "3D + Time Knowledge Chart Diagram" (**Triangle Q**!).

We should structure our education system so that it gets the maximum possible extension of the complexity of thinking for our young ones, compared to their ages. In other words, **instead** of all this funny showbiz stuff in school, (to make our kids **feel good**!) it is time to extend the usage of the capacity of our brain from 5 to 10 %, and maybe **higher**, starting early on in the basic schooling.

Part 2 - Chapter 13
Biology, Psychology

Psychology (cont.)

2. The extended "3D + Time Universe" Knowledge Chart diagram (cont.)

For example: in the USA, 25% has 5th grade reading level, 25% has 8th grade reading level; What about the logical thinking, comprehension, and "3D + Time Knowledge Chart Diagram" of those and their resulting "**Triangle Q?**" And this is **One of the best countries (USA) on the planet**!

The third implication is also from the "Latin education time" of my life, that is the argument was made that we have two different kinds of scientific minds. That observation also fits into the "3D + Time Knowledge Chart Diagram" perfectly. One kind of mind is one that has a very high achievement in **one specific** knowledge area — like biology or mathematics. These are the researchers in the different areas of science.

The other kind of scientific mind is able to **observe and absorb large amounts of knowledge in many different sciences**, but will never reach the level of those specialists, not even come close. (More in Chapter 17!) These people are (myself included) capable of working and understanding "systems" in life without knowing all of the specifics about them. For example, I don't have very detailed knowledge of genetics, to know every base triplets, but I know the basics of genetics better than average.

I know how genetics, as part of biology, fits into the system of sciences and where it belongs in philosophy. Two final conclusions here, **one** that we (humanity) need both scientific minds, and they both are equally important; the **second** is that the specialist will have **one outstanding arrow** in their chart, while the non-specialist will have **more above average arrows and a bigger "Triangle Q"**!

The next comment is that with the proper "3D + Time Knowledge Chart Diagram" we can measure everybody's level of knowledge. (Scary, isn't it? And we could do this with much more precision than the standard IQ test).

It can give us a great indication of the levels of knowledge base for each individual, and also the societies in the planet that are built from those individuals. Why am I hanging on to some what seems-to-be so obvious facts? One can argue that this chart is not a big deal, however, that One missed the whole point here.

Part 2 - Chapter 13
Biology, Psychology

Psychology (cont.)

2. The extended "3D + Time Universe" Knowledge Chart diagram (cont.)

The main importance of the "3D + Time Knowledge Chart Diagram" is that this chart represents the level of complexity and comprehension of one's own mind. The knowledge, experiences, and thought processes — that is the "screen door" that one looks through to view the whole world and every issue in it. For example, if you look at most of the industrialized western world, the individuals in this world strongly believe (large percentage of them) in the higher power that created our universe. For these individuals the theory of evolution is the work of the devil or if they want to be nicer, just all kinds of mumbo-jumbo. Their "screen door" is a "steal door" where there is no room for light to come in or other opinion to be even considered. That is coming from the conditioning of their mind throughout their upbringing.

Only people like myself who had an open-minded upbringing keep the "screen door" open, not replacing it with any solid "steal door" — in other words, they try to keep the mind open for all kinds of philosophies. It does not mean that one has to accept any philosophy (that includes McKaneism), but at least should have an open mind to listen for other thought processes.

My readers need to know, that I was influenced through my upbringing with the Religion of the Roman Catholic Church, the teaching of Socialism, Communism, and through their criticism, Capitalism. Later in my life I lived in and observed Capitalism first hand, and I had a chance to learn about Zen through my martial art practices, and many other philosophies through extensive reading. When one learns a lot about many of these philosophies, one's mind opens up and lets the lights in through one's "screen door". That openness is what most countries education systems try to take away from their young ones. It is a lot easier to deal with simple minds in any society, however, the society creates a more shallow individual to be built from.

That is why I will try to stick with an unedited (other than readable English) and uncensored publication, as far as the thought process itself goes.

157

Part 2 - Chapter 13
Biology, Psychology

Psychology (cont.)

2. The extended "3D + Time Universe" Knowledge Chart diagram (cont.)

That will make this publication, in some places, hard to read and hard to understand, however, that is a minor disadvantage compared to the originality that these thought processes of mine can introduce and deliver to my readers. I want this book to reflect my mind, (in the beginning of the 21st century in the year 2003) my "3D + Time Knowledge Chart Diagram", as I acquired the knowledge through my 30 years part time study of philosophy.

This concept of the measurable chart will have a great implication for our future school systems. They will discover how to teach and test their students, how to check the logical comprehension of their knowledge, and how they can alter their teaching techniques to relate closer and deliver a greater impact on their students' minds. The goal is to find the way with the minimum time spent and the maximum knowledge gained in our schools.

The last very important topic under the extended "3D + Time Knowledge Chart Diagram" is related to our human communication. We (humans) in the daily life communicate every day to accomplish common goals, to exchange thoughts, instructions, beliefs. Basically in short, our communication is revolving around our Level 2 feelings, emotions, and Level 3 Logical / Reasoning thought processes (Chart 11). The situation is simple when we all are working on the same "goal" to achieve. For example, in the work place, all the people know and understand the company mission — that is the driving force of the communications and activities. The complexity begins when people try to debate thoughts and ideas.

A very powerful explanation can be found in the "3D + Time Knowledge Chart Diagram". This 3D chart, as one can see in Chart 29, is one that we all can have, with our own "**3D Object**" to describe our knowledge. If one takes two individuals and charts their 3D chart, one needs to find "**Overlapping Knowledge**" on the charts for the two to be able to communicate with each other. That "**overlapping knowledge**" concept is represented in Chart 30; for simple understanding in a 2D chart on the next page:

Part 2 - Chapter 13
Biology, Psychology

Psychology (cont.)

2. The extended "3D + Time Universe" Knowledge Chart diagram (cont.)

Chart 30

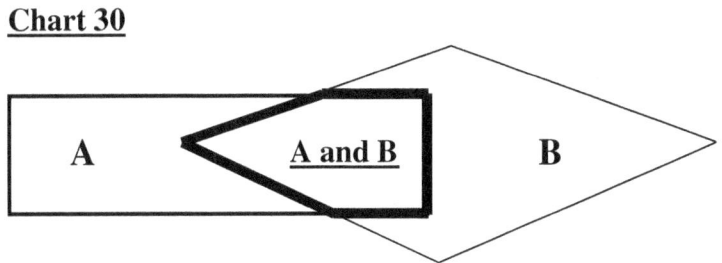

One can see that **A and B** are the overlapping knowledge charts in 2D, to show the common areas of the two individual communicators. (I use a 2D chart here because it is simple enough to represent my argument and I don't have to draw a complicated 3D chart. In reality, it is an "**Overlapping**" "3D + Time Knowledge Chart Diagram"!) Now, one can see how important it is to have overlapping knowledge — it is the existence for human arguments and communication. In today's societies (countries) unfortunately, because of the language problems, (words don't exist with the same meanings in both languages!) or the different religions or different emotional existences, people can't communicate. (Strong argument for a One Language, One Nation society!)

For example: if one follows the Bible and believes the existence of "Heaven" after life, then that individual won't find an "**Overlapping**", "**A and B**" (Chart 30) with one who is based on the Eastern Buddhist belief system, believing one's reincarnation and reappearance on planet Earth. If one draws a "3D + Time Knowledge Chart Diagram", one cannot find any overlapping knowledge of their belief systems in those two individuals' ("3D Object") charts. That can lead to a logical conclusion that the two of those individuals cannot ever work out there differences and they won't understand each others belief system.

Unfortunately, in our current world, religions are so deeply conditioned into people that they won't even consider listening to the other sides' arguments.

Part 2 - Chapter 13
Biology, Psychology

Psychology (cont.)

2. The extended "3D + Time Universe" Knowledge Chart diagram (cont.)

That leads to another very powerful conclusion, that the way we raise our children, we need to introduce them to all the philosophies, at least briefly, and let them pick their own while they at least know a little about all the others.

3. The extended Attributes of Human Beings

We already spent some time in Part 1 to investigate the extended attributes of humanity. I made an argument that only human beings are able to create artificial instruments that can extend our attributes beyond any other species in the natural world. We build microscopes and telescopes to extend our vision to the world of the very small and very large and far objects. We build radio telescopes, so we can "listen" the Universe and its noises from billions of light years and many frequencies, that we can't even hear with our ears, and the list can go on and on.

Why is it important ?

It is important because one of the most accurate realistic pieces of information of our knowledge is coming from observations. We observe our outside and inside reality with our senses. We noticed in the animal world that some species have much more accurate senses than our (humanity) own. However, in the animal world, one outstanding sense, most of the time, means that the other senses are very underdeveloped, or maybe even non-existent. Our extended senses help us understand and comprehend our world better every day.

That also impacts our mind. The extending knowledge base helps us to understand what is around us, near and far. When we create experiments based on our observations, we can prove our theories (that are coming from our thought processes) factual. I will go back to the investigation of our logical mind in Chapter 17. Here I want to focus more on the **"Power and Limitation"** of humanity, even with our extended attributes.

One of our extended and every day extending attributes, that I didn't address yet, is our world of Computer Science. The computer extends the speed of our mind and the capacity to store information.

Part 2 - Chapter 13
Biology, Psychology

Psychology (cont.)

3. The extended Attributes of Human Beings (cont.)

One can look at the Internet and realize that the information that is available on the Internet can't be memorized by one human being in the year 2003. I must define a time parameter (as we see in the "3D + Time Knowledge Chart Diagram", Chart 29) because time changes our statements, proving the world a relative, instead of a static existence. (More about this in Chapter 17.) However, with all these (airplane to overcome distances, microscopes, telescopes, radio telescopes, computers) fantastic extensions, humanity still has our existing boundaries.

We are extending our boundaries with our technology, but we need to understand that we can't get out of them. We can go underwater, and stay there for an extended period of time (scuba diving, submarines) but we don't change our attribute so that we still are land based and need air that we can't obtain under water, however, in these examples our technology extended our capability to stay under water for a longer period of time, than we could do with our born attributes.

One can argue (after all this is a philosophical publication) about how do we change ourselves to be able to live under water, or plant our mind in a robot, so we can stay out in space with no oxygen consumption requirements?! This is a great argument, however, then we have to start arguing about the attributes of robots, because those species won't be humans anymore.

One needs to understand that these definitions are not a playground of philosophy, but a concept to put our thought process in the right place and define and call an apple for an apple and an orange for an orange. The big confusion is coming in the year 2003, that people think if they play with words, that will change the reality around them. So if I don't like my reality, I use another definition with different words and reality will change for me.

I have news for you — It Won't! One needs to remember that observation of the reality around us is the reflection of things and events in our mind. (Mirroring the different formation of matter and events in our Mind). I did hear this "very stupid" statement that said "We all have our own reality." It is absolutely incorrect.

Part 2 - Chapter 13
Biology, Psychology

Psychology (cont.)

3. The extended Attributes of Human Beings (cont.)

Here is a funny example as to why: if I see a rock on the beach, that is my reality. If I drink a lot, I will see two rocks while viewing the same object, so that will be my own "new" reality. However, reality didn't change because in the one moment of time only one existing reality is out there, regardless of my observations.

To prove that, in this situation we can video tape this rock and play the tape back and many viewers can verify the real reality. (If that is not satisfactory, then we can ask many individuals to take pictures of the rock with their camera, and we can go on with this experiment until we can scientifically prove the reality of the "one rock" existence!)

I will revisit in Chapter 17 this relative reasoning also. The correct statement would be that **"We observe the real reality in our own different ways through our own "screen door!"**. That exactly corresponds with the subject of the next paragraph.

4. Some questions about our thinking or thought processes

Next I need to extend on one question that still is unanswered in my mind and maybe this will be a good opportunity for some scientist to use the influence of my philosophy to try to answer it for all of us. Here is the problem. My native language is Hungarian, so English is my second language. (Like you couldn't tell by how "well" this publication is written!) In the beginning of my life in the USA I was learning English as a second language and did catch myself many times translating my Hungarian to English. That created many goofy situations in my life, because some words translated exactly one on one from Hungarian to English, while others had no meaning or were a laughable meaning in English. (For example, "Hippo" is called a "Water Horse" in the direct Hungarian to English word by word translation.)

It sounds like a goofy subject in a serious philosophy book, but this investigation delivered an extremely serious question, that I think is still unanswered in today's science world. **Are we *thinking* through our own native or learned language, or is the *thinking* process of the mind a language independent process?**

Part 2 - Chapter 13
Biology, Psychology

Psychology (cont.)

4. Some questions about our thinking or thought processes (cont.)

If one thinks it is easy to answer, try it. I know from my own experience that some of my logical thinking that I acquired in Hungary didn't change in the USA, however, I replaced the Hungarian words with the English ones.

I will investigate the different logical thinking in Chapter 17. For now, the interesting question — If I am a fully grown adult person, and my logical thinking is set, and I am learning a new language: Am I thinking through my first language with the same independent logic?, or the different new language (new words) is changing my logical thought process? I don't know the answer, but this answer will discover one more mystery of our mind.

5. My philosophical theories of the directions of the self-realizing mind

Here is another very important subject. Humanity has a Self-Realizing, (Mirroring reality) mind. That means that our mind not just registering our environment but also ourselves. Many times before I argued that our human existence is driven by our emotional 2nd Level attributes in the year 2003. The first thing we do almost every morning is to look into our mirrors to see ourselves. That leads us to the first trend in our society (again driven by our emotions) to look the way we like to see ourselves in the mirror. That leads us to more and more popular plastic surgeries in our society. We (humanity) the thinking, feeling minds make a decision about our look and alter ourselves. In our emotions-controlled-and-ran society it isn't a surprise for me.

Then we are going to the next level, that is "Cloning", to experiment with the same genetic equivalent or continuation of ourselves. Of course, that sets up tons of problems, because in the emotional and religious based society only God has the right to alter or create our life. I have bad news for all the religious leaders.

You can't stop a Self-Realizing mind to follow the Directional Theory, (that is discovered and described by McKaneism) to develop many ways to alter our body, mind, and try to extend our limitations in our existence (like a short life).

Part 2 - Chapter 13
Biology, Psychology

Psychology (cont.)

5. My philosophical theories of the directions of the self-realizing mind (cont.)

For now in 2003, the religious leaders can influence the politicians to make some of these processes criminal or illegal, but the future events of humanity will remove them as road blocks of human development. They will be "gone with the wind". If one thinks that this process will stop there, don't bet on it.

The next step will be with the enhancement of robotics and computers, humanity may or will try to mix the best features of robots and humans. The constant search for extending our life span, to more efficiently use our brain, to overcome in our environment, to extend our existence in space, are just the few unstoppable trends that humanity will follow. If you try to stand in the way of the Directional Theory of human development, you will end up the same way if you tried to stand a front of a freight train and raised your hand to stop the train. It won't happen. No "one human individual" can stop the progress of humanity. One's force is nothing compared to the forces of all the people on this planet.

The next, today maybe unimaginable possibility, is that we will be able to replace some of our "broken parts" with new ones, maybe artificial new ones. And one day someone will try to transfer the human conscious mind into a very durable, hard to destruct, long lasting body. These are only a few possible enhancements that one can very easily foresee in the future of humanity. Those steps will be a natural progression of the human mind according to the Directional Theory of McKaneism.

It may sound like science fiction or weird today, but it will be natural and well established in the future — just as flying is today, that was unimaginable three hundred years ago as it exists today. They observed the birds flying, but I don't think anybody in the 15th century could imagine that one can get on an airplane in one continent, and few hours later, arrive in another continent. The Self-Realizing mind keeps improving the Body and Mind, in other words the Matter and Mind Level 1,2,3 existence, (Chart 11) and the environment it exists in, as time goes on.

Part 2 - Chapter 14
Sociology, First Level, The "Fair Share Distribution System Theory"

Sociology, First Level, The "Fair Share Distribution System Theory"

In this chapter I will introduce a "small model" to investigate the different possible structures of the production-distribution environment. I understand that in the "real world" the structure is more complex than my "model", however, you will see later that it doesn't have to be, it is "man made" and made complex by "mankind". The most important point is that I will try to make the representation and analysis of these models introducing the "Fair Share Distribution System Theory" that (one can see) is way more advanced than any existing system. Considering the "Directional Theory" and humanity's drive for "Fairness", in my estimation, it will be the next "Production-Distribution" structure for humanity.

A few of the problems that Humanity has: **One** is, that all these systems, in the past, have just been changing by chance (of course the forces of Directional Theory are working), and humanity never before tried to design any of the previous systems.

Two is, even if I take one's argument that we write our own constitutions, many times in history (like USA) we still end up with one country or one nation's level (that is not a whole humanity!) and those who wrote those constitutions before didn't understand the importance of the "Attributes of Human Being Theory" and didn't bring into the design the Level 3 Logical/Reasoning mind (Chart 11), but instead stayed on a Level 2 emotional driven society.

McKaneism as a Philosophy is the first time in history that I will try to argue about how our social structure should be starting with the "Fair Share Distribution System". (We will be extending the possibilities into the future in the next chapter.) Many or most people don't understand, but we (Humanity) already "used up all our credit to the limit!", and now it is payback time.

Unfortunately, humanity is in real bad shape, (as one will understand by the time they are finished reading this book) and doesn't and won't have an overnight solution. Even if the whole humanity buys into my arguments (looking at the situation with Galileo (350 years), it is not likely to happen!) it could take a few generations to get back on the track. (Currently, we count 25 years as a generation!)

Part 2 - Chapter 14
Sociology, First Level, The "Fair Share Distribution System Theory"

Sociology, First Level, The "Fair Share Distribution System Theory" (cont.)

I will introduce here "small models" of the "Capitalist", "McKaneist" (Fair Share Distribution), and "Communist" distribution systems. These models will reflect very closely the differences between those systems. After reviewing those models I will analyze those systems pointing out the major positive and negative aspects of them. If one is willing to view them as "Concept Models" of production-distribution systems, one took the first step to listen and review the arguments of those concepts.

Make no mistake about it "Capitalism" will change with me or without me, and with all of your or without all of your approval. History teaches us that production-distribution systems have changed in the course of history of humanity. The only question is, are we planning to design the next production-distribution system, using our emotional and logical mind, or let it happen as it did in the previous history of humanity? The importance is very great for humanity. I would like to be optimistic to think that we (humanity) will be smart and prepared for the upcoming challenges and we will travel into a better future, for all of us together.

Let me investigate the following models (Chart 31). In these models I will use an easy to understand example of a company that assembles and delivers coffeemakers (those small table top types) to the distribution centers.

This next chart (Chart 31) compares the "Unequal Unfair" Capitalist, the "Unequal Fair" McKaneist, and the "Equal Fair" Communist distribution systems, representing the main differences between those systems in a "2D chart" format.

Part 2 - Chapter 14
Sociology, First Level, The "Fair Share Distribution System Theory"

Sociology, First Level, The "Fair Share Distribution System Theory" (cont.)

<u>**Chart 31**</u>

Capitalist model	McKaneist model	Communist model
Cost of Business	Cost of Business	Cost of Business
Owners of the Company "Set Amount"	% Owners of the Company	Equal distribution between Employees
Government Taxation	% Distribution between Employees	
Set Amount between Employees (Hourly rate !)		

Let us analyze these charts together. The concept is here that our imaginary company buys the parts of the product (Cost of Business), assembles them, and puts them in a truck and delivers them to a distribution center. I am assuming here that my reader understands our (USA) Manufacturing - Distribution Centers - Retail Outlets flow of our products! I understand that we invented many different formats (like direct order from manufacturing, etc.) of this process, but they are not relevant here to introduce the "Concepts", and yet they also fit into the concept even in a more complex example. (I will let everybody play that out for themselves.)

The "players" in this process are: **The** original parts of the coffeemaker, **The** delivery truck, **The** assembly line and **The** building, and of course **The** human elements who perform the activities of these processes. Now we can investigate, "What are the differences?" in the three "model structures" of our sample manufacturing environment.

Part 2 - Chapter 14
Sociology, First Level, The "Fair Share Distribution System Theory"

Sociology, First Level, The "Fair Share Distribution System Theory" (cont.)

It is easy to see that "these three structures" are also the defining structure of the Production-Distribution environment that can be extended to the "one country" or even the "One Nation of humanity" level. I will make the "Cost of Business" the same in all three models. In Communism there are no individual owners of the company (according to their teaching the state = people own everything). In Capitalism and McKaneism ("Fair Share Distribution System") of course our sample company has owner(s), a President, a Supervisor, an Administrator, a Truck driver, and Assembly line workers. The Communist model has the same other than the owner(s).

Let us "kill" the "Communist model" first (easiest) so that we can simplify our investigation more. You can see from the chart that this model required an equal distribution system. That is very highly idealistic. We saw at the investigation of our human attributes, that we are (individuals) very different. We have a different genetic make up, different needs, and different desires. To force an equal distribution on an unequal needs people seems very idealistic, illogical, and unpractical for me. They were claiming that with the advancement of humanity we would do this kind of sharing willingly, but for me it is another idealistic argument. The president of the company has a greater responsibility than an assembly line worker, so it is very difficult in my mind justifying an equal compensation (paycheck), which is the equivalent of equal distribution.

The originators of this philosophy talked like "when from the fountains of plenty everybody equally taken then we arrived at Communism". One can see it is a great play with the human emotions in those societies where great differences exist between groups of people (they are called "classes"), but in reality people never will desire equal sharing because they have unequal needs. (For example I don't smoke cigarettes, so I don't need the equal share of 10 packs of cigarettes a week, however, I want my two to four cigars a month. I want to see someone push this on one of my friends who doesn't smoke at all. One can come up with a wise argument that you can sell or trade the surplus of your needs, but remember, communism (in theory) is based on the availability of plenty of stuff, so no one wants your extra stuff.)

Part 2 - Chapter 14
Sociology, First Level, The "Fair Share Distribution System Theory"

Sociology, First Level, The "Fair Share Distribution System Theory" (cont.)

So think this over, extend your reading if you please, but I have to draw my conclusion here that the "pure" communist production-distribution system is only a theory and it never can work for humanity. In other words, **the communist production-distribution system is a non-working concept**.

One footnote here, those countries in the year 2003 that we call communist (remember in our western upbringing it is equivalent to calling them evil empires) are not really communist countries by the book. They modified their distribution system with some government controlled priorities, (for example bread is important, make it cheap, high grade whisky is a luxury item, make it expensive etc.) and changed their leadership with force, and are constantly shifting to the direction of dictatorships, then eventual failure. They are not sustainable "social structures" because they are not supporting the needs of the people who live under those structures.

So, only a matter of time before they all will collapse, by inside or outside forces, following of course the "good old Directional Theory". They would have collapsed earlier but replacing them with Capitalism is really not much progress in the year 2003. Something better is cooking here, so keep on reading. Now we can compare the two structures of my models that are left standing. I will assign simple numbers to the process so we can investigate a few scenarios of the production-distribution systems of these two.

The basic differences between the two models is that the Capitalist is what I call a **"Static Distribution Model"**, and the McKaneist is a **"Dynamic Distribution Model"**. The reason the Capitalist is a static model is because the owner(s) who are the capitalist (with the capital) of the business, makes the decision about the distribution. If you remember the "Capital - Something Happens - More Capital" model you need to understand that the capital doesn't get involved until the owner of the capital is ensured that "More Capital" will happen. So in our example let us assume that our model company makes $100,000.00 sales a month. That is a $1.2 million dollar operating company per year. That is a very nice small company in the USA. So let us break down these numbers by months.

Part 2 - Chapter 14
Sociology, First Level, The "Fair Share Distribution System Theory"

Sociology, First Level, The "Fair Share Distribution System Theory" (cont.)

We assume for simplicity that the building and assembly line set up is paid in full. (this way no capital appreciation and any other stuff. We will address this in the next chapter at the Second Level of Society structure, and one will see that it is up to us to have those rules. "Artificial Structure" is depending on our decisions!)

Let us see my Chart 32 on the next page for the continuation of our analysis.

Chart 32

Cost of business:	$20,000.00 for the parts to the assembly
	$ 5,000.00 for truck (and gasoline) and administrative expense.
Owner:	$30,000.00 (what the owner wants to take out or no business!)
President:	$10,000.00 (high paid executive salary)
Administrator:	$ 3,000.00 (including secretarial and accounting functions)
Sales Person	$ 5,000.00 (working on extending product sales in the country)
Truck Driver:	$ 2,000.00 (delivering the final products to distribution center)
Supervisor:	$ 3,000.00 (leading the assembly process)
10 Assembly	(10 x $1,500.00 (Monthly wage)
Workers:	$15,000.00
Corporate Profit:	$ 7,000.00 (Taxed by the government!)

TOTAL: $100,000.00 (monthly)

(I understand that each employee costs about 1.13 to 1.17 times his yearly wage, and also that companies can have more expenses to "eat up" the 7 thousand taxable profit, but remember, this is a concept model, and after our (USA) top of the line economists understand my concept differences, I will be happy to analyze the possible detailed set up of the new system with any of them.)   One can see that the owner(s) will set up this kind of business plan before anything happens and if they find it possible, feasible, and doable, they will start a process and this small company will be born.

Anyone can see the **"Static Distribution Model"** here, because the owner(s) make their decision, how will they want to run the "show". Like it or not, they have the capital so they have the saying. I am talking about they, because in reality, most of the companies have more than one owner. However, I will try to stick with one owner in this small model (for simplicity of the example).

Part 2 - Chapter 14
Sociology, First Level, The "Fair Share Distribution System Theory"

Sociology, First Level, The "Fair Share Distribution System Theory" (cont.)

A few observations here. We saw that if the owner doesn't get his share, this whole business model dies and never gets started.

Next the owner already selected a distribution, that will start this company, but this business model is not prepared for the flexibility of the changes in the economy. The other question arrives that "What is my problem?". It looks like a very well designed, ready to run business model. After the owner pays for the building, the assembly line set up, the truck, and hires all the people for their respective positions, everything is ready to "rock".

Let us set up the McKaneist **"Dynamic Distribution Model"**, then we can analyze the two models functioning in the ever changing economic environment. The same example, so let us assume that our model company makes $100,000.00 sales a month. That is a $1.2 million dollar operating company per year.

Same conditions.

Chart 33

25% Cost of business:	$20,000.00 for the parts to the assembly
	$ 5,000.00 for truck (gasoline) and administrative expense.
30% Owner:	$30,000.00 (what the owner wants to take out or no business!)
10% President:	$10,000.00 (high paid executive salary)
3% Administrator:	$ 3,000.00 (including secretarial and accounting functions)
5% Sales Person	$ 5,000.00 (working on extending product sales in the country)
2% Truck Driver:	$ 2,000.00 (delivering the final products to distribution center)
3% Supervisor:	$ 3,000.00 (leading the assembly process)
	(1.5%; 10 Assembly; 10 x $1,500.00 Monthly wage)
15% Workers:	$15,000.00
7% Corporate Profit:	$ 7,000.00 (Taxed by the government!)
100% TOTAL:	$100,000.00 (monthly)

One can see that I introduced the percentage (%) based compensation. I left the same structure intact to start with, however, one can and will see the huge differences when one puts the dynamic model into the reality of economy, compared to the static structure behavior in the dynamic economic environment.

Part 2 - Chapter 14
Sociology, First Level, The "Fair Share Distribution System Theory"

Sociology, First Level, The "Fair Share Distribution System Theory" (cont.)

I also have to note here that we do not need corporate taxation at all. Taxing in the capitalist society ensures a power control. In our USA we can see the struggle between the **companies** (lobbying with huge sums of money, buying power by contributing huge donations to election time, etc.) and the **government** to control the most affluent (the holder of the capital) and their play ground the "Free Market".

I put the free market in quotes, because in the final stage of capitalism there is no such a thing as a free market. This is a regulated market for the big power game between the "Big Capital Holders" and "Big Brother", as we call our government from time to time. So in the McKaneist model, we can start with the elimination of the corporate taxation and that leads to saving the 7% or $7,000.00 for the corporation.

One can say great mind experiment but what is the point? In these models a lot more information is stored than one can discover at the first glance. In case we take a "snapshot" of the economy, the two models show no difference. In reality, however, the economy is not a one point snap shot, picture taking — instead it is more like a movie, rolling with the camera. What I am trying to say here is that the continuous time element of the economy will make a huge difference in the functioning of these two models. Back to our basics, the time (the events between two static points), will make the McKaneist model far superior to the Capitalist model. (**or I can say: will make Capitalism obsolete!**)

I will investigate two examples, (I'll let you pick the rest for your own examination!) the first is when the company's **sales double (200%,** with the monthly $100,000 that is $200,000), the second is when the company **looses 25%** of it's monthly sales. (about a 1/4 of the $100,000.00 that is about $25,000.00 a month!)

We will see how the two models respond to the changing economic conditions, and I will point out why the capitalist model isn't sustainable in the long run.

Part 2 - Chapter 14
Sociology, First Level, The "Fair Share Distribution System Theory"

Sociology, First Level, The "Fair Share Distribution System Theory" (cont.)

Here are the models with double sales: $200,000.00

Chart 34

		McKaneist model:	Capitalist model:
Cost of business:	25%	$40,000.00	$40,000.00
		$10,000.00	$10,000.00
Owner:	30%	$60,000.00	$30,000.00
President:	10%	$20,000.00	$10,000.00
Administrator:	3%	$ 6,000.00	$ 3,000.00
Sales Person	5%	$10,000.00	$ 5,000.00
Truck Driver:	2%	$ 4,000.00	$ 2,000.00
Supervisor:	3%	$ 6,000.00	$ 3,000.00
Workers: (10)	1.5%	$30,000.00	$15,000.00
Corporate Profit:	7%	$14,000.00	$82,000.00
TOTAL:	100%	$200,000.00	$200,000.00

Here we are, of course, that the double sales will require double productivity. For one month the companies in the USA probably can handle with ease, but how about an extended period in our economy, that would keep this increase going for months. "Piece of cake" as we say here at home in the USA for the McKaneist system. Because it is a "Fair Share Distribution System" the people of the company can sit down for a meeting and they can decide what they want to do. For example, if they want to hire 5 more workers to the assembly line, they can figure it out what is the "**fair**" percentage (%) change that they all can live with. Sounds logical that the 1.5 % of the new 5 workers will change the equation.

I have to pause here for a moment to introduce an interesting fact.

If one enrolls in a small business course in our colleges, one will get introduced to a fun formula of "Capitalist" business. In a joking way the instructor will tell you that: "In business, everybody gets tuned to the **WII FM** radio station". That is to say: "**What's In It For Me?**". From the philosophical viewpoint, this little joke sums up "Capitalism". The "Society of the Individuals!" For me, as an "Individual", there has to be some interest in any event or I am not participating.

Part 2 - Chapter 14
Sociology, First Level, The "Fair Share Distribution System Theory"

Sociology, First Level, The "Fair Share Distribution System Theory" (cont.)

That interest can be financial (as a paycheck) or personal (as a good time in the ball game). I will analyze this later in deep details (you guessed right, in Chapter 17), that this individual mind set will lead to a social structure, where if we disagree on something, (which happens all the time with the interest collision) we vote for the solution, and the more votes win, regardless of the feasibility of the solution. That is the "Emotionally Proper Way" to solve the conflicts, but not the "Realistically Proper Way"!.

Here in the McKaneist "Fair Share Distribution System" we will change the "radio station" first, and that will lead to a totally different view of the conflict later. The new radio station is : **WII FU**. That is : "What's In It For Us?". One can see that voting for a conflict by emotions **can't create a "Fair"** solution for anything. The fact of the process that we have to vote, by itself represents a disagreement, and that has to be solved by the many that enforce their view on the few, even if they are wrong! (Does it remind you of the Inquisition of Galileo?!) In the **WII FU** system the people in our company model can sit down and communicate the problems and solutions in their new situation. (Income doubled!) One can see that the new situation effects them all. Because of the "**Fairness**" of the "**Dynamic**" percentage(%)-based distribution, they all have an interest in their changed environment and company structure.

One can see a system where they all can voice their opinions, and have a short intelligent meeting about the situation. The Owner(s), and President after the meeting can make their intelligent best decision, based on all the facts and arguments, that they did hear from all the players. (If one tries to object, that it is impossible in a large company, with hundreds of thousands of employees, that one is wrong. All companies have an organizational structure (mostly by functionality!) in place, so the company can identify the interested groups of people. It can be a division, or maybe just one small office. One can realize after investigating the "Fair Share, Dynamic" model, that it has great flexibility for adjusting constantly to the ever changing outside "free market" or economic conditions.

In this new concept **WII FU**, one can see the change from a "Society of the **ONE** Individual Interest" to a "Society of the **ALL** Individual's Interests"!

Part 2 - Chapter 14
Sociology, First Level, The "Fair Share Distribution System Theory"

Sociology, First Level, The "Fair Share Distribution System Theory" (cont.)

This system doesn't advocate or favor the elimination of the individuals, only the change of their interest structure. In other words, one can see here that McKaneism keeps the "Unequal" part of the distribution of the capitalist system, but replaces the "Unfair" share part with the "Fair" share part.

Let us investigate one possible (but not the only) solution of the changes in our model company to deal with new (this time positive) market conditions.

Let's see one possible change out the many to keep the **fairness** in the **McKaneist system**. The 5 more workers will bump up the number from 10 to 15. So, we need $15,000.00 (5 people 1.5% / month each, $3,000.00) that is to keep the system going. Here is the question: If you are the owner, president, and the sales person, do you mind to contribute when your share really doubled?

If I would be in those positions (and remember like a sales person I live my life on the $5,000.00 / month income and overnight I start to receive $10,000.00.), I have no problem to give up some, to keep it fair. As long as everybody is happy, the production will accelerate and my money will keep growing. Let us see a changed chart after the owner, president, and sales person contribute to the changes.

Chart 35

	McKaneist model:			McKaneist **new** model:	
Cost of business:	25%	$40,000.00		$40,000.00	25%
		$10,000.00		$10,000.00	
Owner:	**30%**	**$60,000.00**		**$50,000.00**	**25%**
President:	**10%**	**$20,000.00**		**$17,000.00**	**8.5%**
Administrator:	3%	$ 6,000.00		$ 6,000.00	
Sales Person	**5%**	**$10,000.00**		**$ 8,000.00**	**4%**
Truck Driver:	2%	$ 4,000.00		$ 4,000.00	
Supervisor:	3%	$ 6,000.00		$ 6,000.00	
Workers: (10)	**1.5%**	**$30,000.00**	**(15)**	**$45,000.00**	**1.5% x 15**
Corporate Profit:	7%	$14,000.00		$14,000.00	
TOTAL:	100%	$200,000.00		$200,000.00	

Part 2 - Chapter 14
Sociology, First Level, The "Fair Share Distribution System Theory"

Sociology, First Level, The "Fair Share Distribution System Theory" (cont.)

So, the owner gave up 5%, the president 1.5%, and the sales person 1%, and they covered the cost of the 5 new workers. Please note here that the owner's new share is $50,000.00, that is $20,000.00 more than the original $30,000.00 before the production doubled! The president makes an extra $7,000.00 and the sales person an extra $3,000.00 a **Month.** Everybody else takes home double their money every month. One very important fact that this really is a dynamic distribution system. There is many ways to adjust the percentage (%) system, as the economy changes and the people can benefit and work together.

One can have an understanding, that this system is still individually based, still unequal (as it should be to reflect the attributes of humanity!) but extremely **fair** to the people who live by it! Another great opportunity to point out, that McKaneism is just "The Foundation of Humanity". One can see here that this model points out the percentage (%) system for fairness, however, I don't set any specific form or shape for the system that companies should work and live by. That is the duty of the society who implements this system, to figure out how to make the Fair Share Distribution System the "**most fair**". However, I will provide more pointers to think about as we go on. We will see next how the capitalist system scrambles for the profit at surplus time, and how it fails big time when the economy contracts, or turns into recession.

Here we are looking at the capitalist model next. The profit increased overnight, so what is going on? Any of you who ever tried to do your own business, (I am one of those entrepreneur types) experienced first hand that the "Governments", (the friendly Big Brothers?!) always keep their eyes on you. Why am I saying "Governments" instead of "Government"? Because the "City", "County", "State", and "Federal (USA)" governments keep very precise records of your business activities. They recognize instantly that you have a huge (in this example) $82,000.00 dollar profit from your monthly business activities. I think you did guess already that they all want their share of the profit. So the company rushes out to figure out what to do with this newly found fortune. The two major players in this game are the owner(s) and the government(s).

Part 2 - Chapter 14
Sociology, First Level, The "Fair Share Distribution System Theory"

Sociology, First Level, The "Fair Share Distribution System Theory" (cont.)

No one else has the saying in the "Capitalist" structure. **That creates the "Unfair" nature of capitalism.** That is a fact of life that this system becomes the "rich people's game" on the account of the rest.

For example, the owner can make a decision that he wants to run the double production with the same 10 workers. (Sound familiar?) Then the fantastic tax shelters coming into play, using smart accountants, tax attorneys, and sometimes "cooking the books", or even walking the thin line of the law. So the game is on, and at the end, all the governments and the owner get part of the new fortune, and life goes on with the double shift for those workers who don't (or can't) quit this company. Don't believe me?! Check out the corporate environment around you. (Most companies follow this practice, to hire someone for a 40 hours per week job, paying the 40 hours salary, but demanding at least 5 to 10 extra hours of work per week. (That is a 12.5% to 25% free work for them!, Sadly I talk with some of my acquaintances who thinks it is OK!?)

Here is the double sales: $200,000.00 for the Capitalist model:

Chart 36

Capitalist model

Cost of business:	$40,000.00
	$10,000.00
Owner:	$30,000.00
President:	$10,000.00
Administrator:	$ 3,000.00
Sales Person	$ 5,000.00
Truck Driver:	$ 2,000.00
Supervisor:	$ 3,000.00
Workers: (10)	$15,000.00
Corporate Profit:	**$82,000.00**
TOTAL:	$200,000.00

One can say, no problem, works well, the owner deserves the big bucks for the risk taking. Here the problem is that in the process of who gets what, the owners and governments power play leads to hundreds or thousands of rules, (7 million+ word tax code, USA).

Part 2 - Chapter 14
Sociology, First Level, The "Fair Share Distribution System Theory"

Sociology, First Level, The "Fair Share Distribution System Theory" (cont.)

That affects the whole system in a negative way. In the process to "Out Fox" each other the two sides slowly create an impossible environment to operate in. Ask any small business owner how they have a harder time every year to conduct their honest business.

The previous example was the better times for Capitalism, now we will see what happens when the economy shrinks about 25% for our sample company.

Here is the 25% reduced sales of $100,000.00 to $75,000.00

Chart 37

			McKaneist model:	Capitalist model:
Cost of business:	25%	$15,000.00		$15,000.00
		$ 3,750.00		$ 3,750.00
Owner:	30%	$22,500.00		$30,000.00
President:	10%	$ 7,500.00		$10,000.00
Administrator:	3%	$ 2,250.00		$ 3,000.00
Sales Person	5%	$ 3,750.00		$ 5,000.00
Truck Driver:	2%	$ 1,500.00		$ 2,000.00
Supervisor:	3%	$ 2,250.00		$ 3,000.00
Workers: (10)	1.5%	$11,250.00		$15,000.00
Corporate Profit:	7%	$ 5,250.00		**-$11,750.00**
TOTAL:	100%	$ 75,000.00		$ 75,000.00

One can see that the McKaneist, "Fair Share Distribution" system goes on. Everybody ends up with some pay cuts, but the company moves on. No one has to lose their job, and most people with sensible money management can go on for a longer time on less money. For one, the workers who get hit the hardest, the monthly pay is reduced from $1,500.00 to $1,125.00. That could mean a change from a new car to an older paid-off one, or cutting back on other items, but life goes on without losing anybodies entire income.

On the other end, our capitalist system is in deep trouble. The -$11,750.00 net loss can't be worked out without changes in the company. Usually the owners don't take any cuts. (What can one expect from an owner who has 12 castles and needs to buy the 13th one?!)

Part 2 - Chapter 14
Sociology, First Level, The "Fair Share Distribution System Theory"

Sociology, First Level, The "Fair Share Distribution System Theory" (cont.)

A possible change is to replace the higher paid people with lower paid ones (president, sales person, administrator) or start cutting the number of workers. As the "Unfair Share" distribution progresses, some people end up in a bad situation. We all know the response for this situation in the USA. Lay offs and more Lay offs. If the economy keeps up in recession, some people can lose their livelihood, even everything that one worked for a lifetime, permanently.

One can see the strength of the McKaneist system, that it has a certain flexibility by it's design. It can tolerate more or less fluctuations of the economic system. If an owner educates their employees for proper money management, one can see that the McKaneist system can go on without major damage in their life, adjusting to the fluctuation of the economy. One needs to understand that the nature of the "Free Market", Supply/Demand, Production-Distribution system, is the fluctuation in time.

One can't say the same thing about the Capitalist (model) system. One needs to understand that because the Capitalist system is built on the Emotional Level 2 observations and reactions of humanity, that system is "hope", "unlimited", "liability" and "freedom" based. All these words try describe "absolute" human emotions and disregard their relative reality. As we saw in our example, the Capitalist system has to exclude individuals to preserve the functioning of the whole system. When the company ends up in a negative number financially, they have to make a hard core decision, who gets out of this production-distribution system cycle?

One can see in this small scale model that the company can't operate in the negative profit (net loss) forever. Some companies do so in today's reality, borrowing money from banks and **hoping** that the economy turns around, so they can pay back the loan (**liability**) and everything can get back to normal. However, this strategy of success depends on the economy. If the negative economic times drag on, the company can't pay back the loan and goes under. (Bankruptcy!) Other companys' owners may choose to dissolve the company, not even trying to keep it going. One can see the mountains of problems that the Capitalist model encounters.

Part 2 - Chapter 14
Sociology, First Level, The "Fair Share Distribution System Theory"

Sociology, First Level, The "Fair Share Distribution System Theory" (cont.)

One also can come up with the argument that it is healthy because just like in the "**Natural System**", the strong survive and the weak fail. However, that one needs to understand that in nature, the weak have no saying in the "brutal" system. In the "**Artificial Human**" system, if you push too many weak aside, they can team up and outnumber the strong ones (became the strong ones!), then they can change the structure of the society! (e.g. French revolution 1789!)

When one understands the working human mind, one can understand the difference between the animal world (passive players) and the human world (active players)! (Example: The small newborn zebra has no chance against a group of hyenas when discovered (passive player), even the mother can't save the young one without risking her own life. In the human world the suppressed ones, when their number became bigger than the suppressors, they could invent means and ways to outsmart the suppressors (active players), using guns, explosives, surprise attacks, etc. Because of the "Human Level 2 / Level 3" mind (Chart 11), the man made system isn't a "Natural System!")

It looks like the "Capitalist" small model has a great amount of difficulty adjusting to the changes of the market conditions. The "Fair Share Distribution System" small model doesn't! One can argue that these are small hypothetical models, however, that one doesn't understand the concept and the difference of these three sample "small models" — if one doesn't see the difference from the viewpoint of "Unfair vs. Fair" and "Static vs. Dynamic" behavior of those systems.

Finally, I will introduce a few more "Attributes" that have to exist to make the "Fair Share Distribution System" a viable working solution for the society. Here I will base my arguments on some facts that anyone can challenge, because I won't argue about them here, but I will revisit them later. Here, only the existence of them is important, and the lengthy writing of the proof of my correctness (that I will do in Chapter 17) would sidetrack the importance of following very important conclusions, and of course like always they are (realistic) observation based.

Part 2 - Chapter 14
Sociology, First Level, The "Fair Share Distribution System Theory"

Sociology, First Level, The "Fair Share Distribution System Theory" (cont.)

"Capitalism" would work well if our environment "Planet Earth" was not a "**Closed System**". One needs to understand that we have a "**Measurable**" size of landmass (to exist or farm on), natural resources (oil, gas, lumber, etc., to use for our purposes), and live stock (to consume, on the ranch or in nature). All of those things can support only a certain amount of human life in the "good existence level" on this planet. So, the idea in Capitalism that we all can have the billions, is false! Even the statement that "anybody" can make it rich with hard work is incorrect. In one point of time in history we (humanity) had a well defined, finite set of wealth for a well defined number of people. The argument is correct that the wealth of the society increases in time, however, the number of individuals also increases in time.

More than that, the wealth increase of the societies is limited by a finite number in the "Closed System"! (more proof about this later!) The same conditions exist for the "Fair Share Distribution System" also, however, in the "Fair Share Distribution System" (because it is Level 3 logically designed) one can realize additional actions from the society that can be taken to make it work. So far, I just started the design, more elements need to be addressed.

The society needs to limit the maximum wealth that each individual can have! It is easy to see that, for example, if one country has it all, then the rest of the world will hate that country for it. Or if I have all the billions of dollars, then that leaves nothing for the rest in the society. That doesn't mean that I try to take away anything from anybody, but solving this issue is one of the key items to create a working "Artificial Structure" for any society. (more about this later also!) One can see that the "Fairness" of the distribution system can energize people to participate in this better Production-Distribution environment.

The society needs to eliminate the mindless, social engineering taxations and must replace it with a "Fair" one! One can see that in the "Fair Share Distribution System" by taxing every working individual, the government can raise enough money to support the main needs of the society — "Health, Retirement, Defense"! It is also easy to see that to be fair you can't have a society where one can live on the others hard work.

Part 2 - Chapter 14
Sociology, First Level, The "Fair Share Distribution System Theory"

Sociology, First Level, The "Fair Share Distribution System Theory" (cont.)

So "All" need to be part of the "Distribution" side!, and only those who can be exempt from the "Production" side, are those who can afford it! (for example, kids supported by parents or financially independent individuals!) Don't make up your mind until you read the whole proposal!

One could have suspected that the change of the distribution system would change the relations of everything else in the society. Now, one can understand why philosophers with new ideas were running into the existing power structure throughout the history of mankind! One needs to understand that this change is coming by the forces of the "Directional Theory", so we (humanity) can make it smooth or hectic, whatever we choose, but we can't stop it!

Part 2 - Chapter 15
Sociology, Second Level, Economy, Law, Ethics, Morale, Aesthetics,Religions, Art, Politics, Education

Sociology, Second Level, Economy, Law, Ethics, Morale, Aesthetics, Religions, Art, Politics, Education

In this chapter I will cover the major topics of the Second Level of Sociology (Chart 16). I argue in more detail in this chapters and here I will point out a few more observations. I will stay on the ground of philosophy and from that ground I will extend my investigation of this topic as the continuation of the investigations of the previous chapters.

We visited Sociology, Second Level in Chapter 7, so now I will investigate it from the point of view of an "Unequal Unfair" distribution of capitalism. One read my argument in the previous Chapter 14, that the base is the "Unequal Unfair" production-distribution system of capitalism as a First Level, and these relations at the Second Level are based on the first level. (See Chart 16 to refresh the structure). I will spend some time to argue about these observations.

When individual minds form a social structure, they interact with each other. I argued earlier the most important interaction is the production-distribution system set up. However, the human mind, (and probably any other intelligent mind in this level of existence) is more complex than we can describe in the production-distribution activities. I can argue that the production-distribution activity may take up 8, 10 hours daily in our modern society in the USA, so we have time left for more than only entertainment.

Our inquiring mind will reach out to many interest areas that cover some of the subjects like **Art, Religion, Morale, and Aesthetics**. Some others like **Economy, Law, Politics, and Education** are coming from the conflicts of the ever changing (but in principal) production-distribution environment. I argued before that in capitalism "The Capital" controls the activities of the society. Because the distribution of the capital is "Unequal Unfair" it is easy to see that the same applies to all these areas also. I will provide a few examples here and will revisit some of the areas in Chapter 17.

Art in the current societies is supported by donations, if they are not industrial art. **Aesthetics** is the science of the art, however, in the "Capitalist" society, it is coming into the picture here, because most products and services have to be aesthetically pleasing to our emotional make up to be a best seller.

Part 2 - Chapter 15
Sociology, Second Level, Economy, Law, Ethics, Morale, Aesthetics, Religions, Art, Politics, Education

Sociology, Second Level, Economy, Law, Ethics, Morale, Aesthetics, Religions, Art, Politics, Education (cont.)

So, "**industrial art and industrial aesthetics**" support production, the "**pure art and scientific aesthetics**" are living on the mercy of a few bright and kind individuals whose donations make it possible for their existence. They are extremely important functions of the human mind, but in capitalism if they don't produce "More Capital" (remember for the formula: "Capital - Something Happens - More Capital") they are not important. For example, a design of a building that involves art and aesthetics, other than the science of mechanics and engineering, that is what I call "industrial art".

One who expresses their own state of mind with creating a painting or sculpture is a pure art and the science that investigates the mind functions in that art environment is pure aesthetics. One can easily see that the "Unequal Unfair" distribution system holds the "pure art and aesthetics" in very little value, a more pragmatic "cash" driven society!

Religions and their influence on the **Moral** code of humanity do not do much better either. We have many religions (mostly developed through human history) and they are all exclusive for all but their own members.

Great way to divide society and build tension. The concept of "freedom of religions" was developed by the founding fathers of the USA, but even they couldn't foresee the situation today that religions, the fate based emotions, can drive groups of individuals to collide in the societies. I don't have to go far to point out that their inclusiveness to their own members and exclusiveness to all others is a perfect reflection of the "Unequal Unfair" distribution system. It is raising an extremely interesting logical question.

Why does the "All Mighty" and **fair** creator want to exclude anybody? For example, if you are Muslim, the Christians are infidels, (excluded). If you are Buddhist, the rest are all stupid because they don't go to heaven and the Buddhist knows, that they all will return here through reincarnation, (so much for the 21 virgins that I was personally counting on!), etc. The answer is simple, and I will investigate it further, that all of these systems are man made and fate based.

Part 2 - Chapter 15
Sociology, Second Level, Economy, Law, Ethics, Morale, Aesthetics, Religions, Art, Politics, Education

Sociology, Second Level, Economy, Law, Ethics, Morale, Aesthetics, Religions, Art, Politics, Education (cont.)

They are reflecting the nature of the distribution system, that if you have a bigger group around you, that group can grab a bigger share of the "Unequal Unfair" distribution in the society. One can observe this by visiting the number of their churches and observing the size of their membership, the wealth of their money making publications, etc.

We don't do much better in the area of **Economy, Law, Politics, or Education** either. One can see when it comes to **Education** the resources of one individual defines the level of education that one can receive. (Anyone can come up some exceptions!) Generally, the more money spent on education, then this delivers a higher level of state of mind.

That will correspond later in life most of the time to a bigger share out of the distribution system, not necessarily with a bigger contribution to the system. Small effort on production, great reward on the distribution side. We all did see examples in our life in the USA that the **Law** is different for the rich and the poor, from time to time. It is not suppose to be that way, it should be equal or least "Fair" for all, however, needless to say that the Law of the land also reflects the distribution system — that is "Unequal Unfair". More money buys more influence, better representation, and a different outcome of the legal situation.

Economy is a very interesting area. In the year 2003, we have some extremely intelligent economists in our (USA) world. With the help of their computer modeling and the exact number of production measured by their currency, (dollar or euro), they can forecast, describe, and analyze almost every event in the world of economy. However, their place in our societies now exist only as advisors. It is sad to see that a realistic science, based on logical modeling and hard core measurement and comparisons of events, is simply pushed aside by the next subject matter — the actions of **Politics**. Before I make my argument on the subject of politics, one needs to understand that, for example, the number of around 1 to 2.5 billion people would live a comfortable existence on this planet is coming from the work of economists and biologists (in short, scientists!).

Part 2 - Chapter 15
Sociology, Second Level, Economy, Law, Ethics, Morale, Aesthetics, Religions, Art, Politics, Education

Sociology, Second Level, Economy, Law, Ethics, Morale, Aesthetics, Religions, Art, Politics, Education (cont.)

Politics the science of manipulating the audience for the politicians direction through emotions, triggered by flowery words. Our current interesting invention (of course, it isn't new in the history of humanity, example: Cicero in the Roman Empire) is the speech writers and communication directors, who are spokespeople around each and every politician in our current 21st century. Looks like it works — manipulating the people's opinions to the politician direction, by making them emotionally tuned into some of the issues, using the mass emotional weaknesses. For example, when the issue arrives that the politician cares about the future of our children, which one of us shouldn't like the speech, especially any who has their own children? Sometimes, I question their sincerity in their arguments however. Some of them believe in the goodness of his or her functions, but some of them just like the power, and will do or say anything to get elected.

Here I need to address what has been many times spelled out as big achievements of ours, the "Great Democracy" as a social governing body. We concluded that the "Unequal Unfair" production-distribution system of Capitalism doesn't work. For many of your surprise, about 2500 years ago Aristotle analyzed the structures of societies and concluded that Democracy doesn't work. Here we are two and a half thousand years later and still pushing democracy into a society that was intentionally founded by the "great founding fathers" (USA) to be a Republic. Of course, we like to hear ten times a day that we are a great democracy because again, for our Level 2 emotional based system, it "feels very good". That means that "we the people" have the power.

Or, Do We?

The answer is NO!

Before I investigate these issues and make my arguments, I have to repeat myself that the USA is one of the most advanced societies on the planet, while many other countries are living under a great confusion about their non-working systems, and they aren't even worth analyzing.

Part 2 - Chapter 15
Sociology, Second Level, Economy, Law, Ethics, Morale, Aesthetics, Religions, Art, Politics, Education

Sociology, Second Level, Economy, Law, Ethics, Morale, Aesthetics,

Religions, Art, Politics, Education (cont.)

My arguments here are coming from my observation of philosophy, with my honest intention to introduce thought processes that may or can lead to make this system better, and not from the angle to be critical and trying to put this system down.

Even the best manmade system has room for improvements! Some of my preceding philosophers, like Marx, introduced the class definition, and unfortunately, it still exists in our modern language. In the USA we are talking "High Class", "Middle Class" and nicely "Lower Class" or not so nicely "Low Class", which is an insult mostly on anybody who named there. The definition coming from the measurement of financial wealth of the individual, and what kind of lifestyle one can create on that certain level of wealth.

I will argue later that the class theories, and the class fighting each other theories, greatly misleads humanity from the most important challenges of humanity. Somehow we have a belief that if the middle and low class control the power through voting for their leaders, then the great representative democracy is working. Here are a few problems.

Even in the advanced society like the USA, about 120-130 million households file their tax returns, yet the vote count is running around 51 million. (I think the number was somewhere around there in 2000). So significantly less than half of the people are interested in participating in this voting type of society.

One can ask: are these people not interested about their leaders and their policies? The answer is that these people are smart people and they realized that a long time ago, that voting is a waste of time. Look closer at who is running for the big offices. People with many millions to be able to finance their campaign. Have you ever seen "Joe Blow the painter" running for president? Of course not, it isn't even an option. So where are the nominations coming from? Big surprise? No, they are coming from the high class, big money!

One can argue, so what? If they are good moral characters, with good plans to serve the people, then what is the problem. Here are a few problems with this system.

Part 2 - Chapter 15
Sociology, Second Level, Economy, Law, Ethics, Morale, Aesthetics, Religions, Art, Politics, Education

Sociology, Second Level, Economy, Law, Ethics, Morale, Aesthetics, Religions, Art, Politics, Education (cont.)

First of all the individual who votes gives away his control over the society for the time until the next voting is coming. Based on what those candidates said, one elects them into the office, and **trusting** them, (based on emotional words instead of realistic, factual actions!) to make good laws for the land. Some of those leaders learned the manipulation of the people's emotions, and of course with the help of the media, the press, they managed to sit in their office for decades.

I am questioning their intentions and performance, based on the intention of the founding fathers who designed this system to keep rotating those leaders in and out, and keep the process dynamic. However, these leaders continue sitting in their office, voting for their own pay raise, voting many times against their term limit amendment — so they like to sit there within the "beltway" (USA), where they never get held responsible for their votes. Their performance evaluation didn't get tied to the well being of "We the people" where they came from.

The economy goes good or bad, and they sit there making themselves get re-elected, and sitting in the spotlight of the media, doing very little good for the people. Here is one recommendation, how about they get a paycheck for whatever the current per capita income is, about $21,000.00 in 2003? They may have a raise anytime where they bump up the per capita income higher with their smart legislation. How about the same 401K that "we the people" have and the same "health care mess" that we the people are in?

Let's see how many of them will hang around next election because they are patriotic, and how many of them will cash in their million dollars pension plan and disappear in the shadows of life. The point is that they have a rich peoples game, fame and fortune, and in that atmosphere, it is very hard to remember for us "we the people"!

The problem with the "class warfare" is that only different individuals get access to the leadership positions, and they get just as spoiled as the others were before them. The solution is a system change, not class warfare.

Part 2 - Chapter 15
Sociology, Second Level, Economy, Law, Ethics, Morale, Aesthetics, Religions, Art, Politics, Education

Sociology, Second Level, Economy, Law, Ethics, Morale, Aesthetics,

Religions, Art, Politics, Education (cont.)

One needs to introduce a system (I hope I am helping here), where the checks and balances apply to every "player", and nobody gets exempted or moved above the law. I am trying to introduce a better system, that could be the base and starting point for the people who actually live in it.

All these problems are nicely packaged under the kind word of **Politics**. Here is another thought as to why the Marxism class warfare doesn't work. For example, our (USA) top wealthy high class (about 400 individuals) hold about 880 billion dollars of wealth. (Mostly in working corporate and private capital!) That seems like a big number.

However, if in the name of the "class warfare" one divides that wealth "equally" and "fairly" between 300 million people (About the population of USA), one can realize that we all would end up with one time pay of about \$2,934.00 dollars. (rounded) So, in other words, we are back to my original argument, that we all will be broke, and we won't have working capital in anybodies hand. That is exactly what the socialist/communist countries did after WWII, and they were surprised that they ended up in a state run, broke society. I hope one is starting to see that the solution isn't as easy as that one thinks.

I made a statement in the beginning that McKaneism won't be the exception from the analysis and criticism. I have to point out here the weakness of the "Fair Share Distribution System" and that is the population explosion.

One needs to realize that it is impossible to design or naturally move into a system in a "Closed Environment" like planet Earth, with an ever growing, out of control population! The "Fair Share Distribution System", as an interval, will solve the problem for awhile, however, we need to reduce the planet Earth's population peacefully, or extend our technology to be able to leave our planet and start populating others with our excess population. One needs to understand that I am proposing a peaceful education of the planet Earth to think about the possibilities of reducing the population peacefully.

Part 2 - Chapter 15
Sociology, Second Level, Economy, Law, Ethics, Morale, Aesthetics, Religions, Art, Politics, Education

Sociology, Second Level, Economy, Law, Ethics, Morale, Aesthetics, Religions, Art, Politics, Education (cont.)

One also has to realize that the model of the "Fair Share Distribution System" worked well in the fluctuating economy, a lot better and longer than our current system, however, the number of the exploding population can throw that system out of balance also after a longer period of time.

It is time to see the new "Concept" of existence for humanity, that I will introduce and investigate in the next chapter. We need a "New Way to Look our Existence and our Future", if we (humanity) want to have a future. However, an "Interval" existing "Fair Share Distribution System" could address and cure most of the problems of our existing systems for the time being.

We need to realize that it is a fact, that we have to keep our population in balance with the number that our planet Earth can support "comfortably". We can ask our best scientists to come up with some possible studies and numbers. If one thinks it is a "stupid subject of McKaneism", I have to remind that one that not addressing this issue can drag our planet Earth into a miserable existence on this planet for all. Constant fighting for survival, fighting for which religion's "God" is supreme, exhaustion of the resources, self-destruction or each-other destruction on the national level, all is the final destruction of the human race.

If this isn't a "Homo sapiens" planet security interest, then you tell me what it is?! One needs to understand that an uninhabitable planet can't be home for anyone, regardless of who survives the other. I will investigate this subject with a lot more in future chapters, but before I go there, let us spend some time with the "Interaction of Intelligent Species Theory".

Part 2 - Chapter 16
Interaction of Intelligent Species Theory, The Future of Humanity

Interaction of Intelligent Species Theory

McKaneism as a philosophy wouldn't be complete without the investigation of the possible interaction of Humanity and other Intelligent Existence. One needs to understand that the next step on the complexity of the "Matter and Mind" connection is the "Interaction of the Intelligent Species Theory". Like many times, because McKaneism is a realistic philosophy, this part of the investigation will stay on the theoretical level.

Unfortunately, because of the governments of the human race not serving the people, only serving their leading groups, (living on the back of the people, Chapter 14, 15) one can see that the governments and the leading groups, (rich corporate leaders, leading church leaders, leading government officials, etc.) hide all the information that may or may not be known about other existences. One can imagine the impact on "Planet Earth" if we can run into an alien race that traveled billions of light years to get here, and would tell us that no God is "out there". In our planet, of the 95% believers, this news can have a devastating and destroying impact on the societies. So we are kept in the dark as far as any sign of other existences "out there".

Of course it can't stop a reasoning mind to investigate the possibilities of that kind of encounter. There is a lot of speculation and very little facts on this subject. One needs to understand that our "Intelligence" started to form about a "short" (in the scale of the "universe") 5 million years ago by evolution, by chance. (Here again! We find human remains way back but we don't have any evidence for the Creation Theory!) So I go with what we "Know" instead what we "Believe" in the year 2003. No encounters so far. If one wants to prove alien existence, bring one representative of them over to dinner at my place, I'll buy the dinner!

In reality the chance for intelligent life existence, and to encounter them, is very narrow. One needs to understand that the distances and time between possible "solar systems" that can carry life is very large. We can talk about 10 or 100 thousand or millions or billions of light years. Some of these distances are so large that the human mind isn't even able to comprehend.

Part 2 - Chapter 16
Interaction of Intelligent Species Theory, The Future of Humanity

Interaction of Intelligent Species Theory (cont.)

In our current scientific knowledge we need to accept Einstein's theory about the maximum possible speed to travel is the speed of light, as of today's science. (For me it is hard to accept anything absolute in a relative world, but I'll do it for now! I will argue on the relative matter and mind existence concept in Chapter 17.) For now, we simply can't comprehend numbers that our "Milky Way Galaxy" is 100,000 light years in diameter. We understand that is big, but "how big is big" — we don't have any life experience near that size. The same holds true for the number that this galaxy contains an estimated 100 millions or billions of stars such as our Sun. The number of possible planets are "countless".

However, the conditions that intelligent life form, develop, exist, and send out a signal that we can receive, is very slim. Anything we receive has happened in the past. If we receive something from the distance of 10,000 light years, that signal was sent 10,000 years ago. That is a huge amount of time (relatively speaking), and anything can happen with another existence in that time interval while their signal was travelling.

The next problem is that periodically, in the huge universe, catastrophic events happen, like the extinction of the dinosaurs. That can happen with any existence, so by the time we receive a signal, the senders may not be there anymore, or we won't be here to receive it.

We also can have a problem that some of the more advanced species don't want to contact us. (I, for one, sure avoid planet Earth 2003, if I can have the possibility to do so!) Can you imagine an "Intelligent Existence" arriving from a few billion light years away, then finding itself with an argument with our "Churches", to argue about the existence of God, that they didn't experience in their billions of years journey! More than that, to put up with our competing governments, as to who should get the "new alien technology" to wipe out the rest of the planet's existing countries! I am personally sure that I don't want to be that alien existence to encounter this totally messed up "Planet Earth", with all the messed up human minds in 2003!

Basically and sadly (another emotional status) we are stuck in our "cradle", "Planet Earth" for now, by ourselves alone. (More about this later.)

Part 2 - Chapter 16
Interaction of Intelligent Species Theory, The Future of Humanity

Interaction of Intelligent Species Theory (cont.)

One more note, as I see ourselves as an "intelligent existence?", we are "**very arrogant in our existence**". We use our resources as we please and care very little for other existences. We extinct plant and animal life forms as we please. That can lead to an important observation, that if we in the future encounter a technically more advanced civilization, they may choose to extinct us as we did with some of the animal species, or as we are about to do to them (whales, the constant appetite for whaling).

The other side of the coin is what will we do to them if they happen to be less advanced and we (humanity) will be more advanced? I think we need to think about these issues before we run into the situation, with no plans. If we encounter an equal, advanced society in the future, and the two of a kind of "intelligence?" meet, it is hard to imagine a peaceful, friendly relationship between the parties.

This encounter can happen sooner or later, because as a logical conclusion, that is the next level of Matter and Mind interaction — in the complexity level, that will be a contact between two intelligent existences. Without moving into science fiction, we need to understand that the interest of humanity is to be prepared, for that time when and if it comes. It is hard for me to imagine how we (humanity) will be behaving, if we can't sort out our own differences and problems between humans as of the year 2003.

The positive thing is that it looks like we have a good amount of time to go, because our fate and emotion based societies don't push too hard for the search of the other kind. It will be devastating for the religions, to find out that we encountered an intelligence that traveled billions of light years in space and didn't find the existence of any God or Gods at all. Of course our fate based societies simply won't believe them. (Another emotional based decision!) Here, instead of going into science fiction, I will move on to the next topic, and I will try to introduce a totally different concept of existence for humanity for the future. It won't be science fiction, my arguments will be based on the observations of the environment and ourselves.

Part 2 - Chapter 16
Interaction of Intelligent Species Theory, The Future of Humanity

The Future of Humanity

The Future of Humanity. Do we have any? Sadly, as we go today the answer **may** well be **NO**! Of course, that is the fantastic positive aspect of humanity, at times when existence is in question, we reinvent ourselves, and find many solutions for our problems. I will make an argument that another reason why Philosophy is extremely important, because Philosophy leads the new way. By the time someone has some "little knowledge of philosophy", the person changes a lot. (For those of you who try to hang on to this sentence of mine, a "little knowledge", wait until you finish reading this book and then look into your mirror!)

These changes lead to self realization of oneself and the surrounding environment in the grand scale. To think about the future of humanity, one needs to care about it. One who reads this book can see that I am not without a lot of different emotions when I am writing this book. I am worried and care a lot about the future of humanity and my home the USA. I want to see a strong nation like we are (USA) to lead the world into this next millennium, but not by bullying it like the biggest kid on the block, but by peaceful, strong, convincing arguments, as a leader, that other nations can't ignore. Instead of trying to spread the non-working Capitalism, Democracy, and the most exclusive religions, we need to show the world a better solution, or the others on this planet will figure out something better and we will go down like the "Mighty Roman Empire" did.

Let's spend some time with our own reality. We individuals are born into a kingdom of humanity, without having any say about it. Our parents decide they want a family and then, we arrive. I made an argument before that 70, 80% of us (Humanity) are living in some environment that isn't worth living in. Here is the measurement for all of you.

Does your environment allow all the individuals to grow up and function in the average 21st century setting? The answer is **NO**! The average here (USA) is the average of the well developed industrialized nations, and there are about a dozen of them, as I said this earlier. Let us continue in their "better" world. The first six years in our life is mostly wasted, as a development. For example, the brain size doubles between ages 3 and 6 — maybe one good time for "information load?" because it is capable of absorbing a lot.

Part 2 - Chapter 16
Interaction of Intelligent Species Theory, The Future of Humanity

The Future of Humanity (cont.)

Then at age six we start to close our young one's mind, and start to train them within the "OK corral". What I mean is that we start to build the limitation around their mind. They pick up some very basic science knowledge, and a very large amount of emotional enforcement. Very little Level 3 development and a whole lot of Level 2 development (Chart 11). Constantly try to make them "feel good" in school!, then with one of the exclusive religions, we set them up in a one directional and one dimensional thinking existence.

Around the age of 18, some of them have a chance to go to college, to continue their education, where they pick up at least a little more Level 3 Logical development. Then they can go to work for about 40 years, then they can wait for the end in some staging environment. (retirement community or nursing home?) Of course, it is all OK, because it is only a "temporary existence", getting ready for the "forever existence".

Is there anybody home?

How about **No** "forever existence"? How about we should get the best deal out of this one, because this is it!

How about "it has to be something after life belief" because otherwise our Level 2 Emotional mindset, education, and our weak personal make-up will self-destruct us? How about we are too weak to live with the thought that this is it. Enjoy it, because that is all we got, about 75 to 100 years of average "interval" existence on planet Earth. It sounds so scary?, that we have to have "Heaven" for the good ones, and of course "Hell" for the bad ones, or we have to come back through reincarnation to exist again?

It is almost unbelievable for our forcefully developed Level 2 Emotional mind, that maybe we should start developing the Level 3 Logical side more actively and find some other way to look, design, and structure our (humanity) existence. I will try to present to you here, how we can look at our existence in another perspective (through an open minded "screen door"). It has been from the beginning for "one individual", that the "forces of nature" and the "forces of the society" were always overwhelming.

Part 2 - Chapter 16
Interaction of Intelligent Species Theory, The Future of Humanity

The Future of Humanity (cont.)

Then we started to gain very small control over some of the "forces of nature", that we find overwhelming. (climate control in our homes, regardless of the outside weather!) And we invented that in that huge non-controlled environment that it has to be a great higher power "Who" can control the rest of the universe. Very bad thinking in the wrong direction. (or non-thinking!)

Picture the force of an A-bomb that can kill 10 million people in a city in a few seconds, and we (humanity) think proudly about our own huge force. That is not even "baby play" in the scale of the universe. Picture two Earth-sized planets colliding while traveling within a quarter of the speed of light, (that is big stuff) and yet that isn't a "baby play" either on the scale of the universe.

Our sun turns into Supernova, that is the beginning of the events that may count for a "baby play" in the scale of the "forces of the universe". The real "adult play" is when two galaxies have a collision with their 100 million stars (like our Sun) each. I hope that one is starting to get my hint. We need to grow up, look out to our universe, and understand that a lot of "huge power" is out there (not higher power, but bigger power!).

The future of Humanity has to be to use our natural resources and our Mind (Mind of Humanity) to gain control over all the powers of the universe, step by step.

Sounds ridiculous, just like the first time when one said we will fly with many times the speed of sound, or we will travel in space. We (humanity) are able to do big tasks, if we have the proper focus, and will understand a direction defined for all of us. In the next few pages I will start the process to outline the first few needed actions to start leading humanity into the direction of the future. (as I said many times before I can't do it for the "whole humanity", but I can start the thinking process with the hope that others will follow!)

Make no mistake about it, this isn't science fiction, we need to start changing fast, before we encounter a powerful event that we can't match. I am not investigating these possibilities to use the Level 2 Emotions of "Scared State of Mind", (which is one of the most powerful for humanity), but to introduce a possible Level 3 Logical / Reasoning solution for our challenges that lie ahead of us.

Part 2 - Chapter 16
Interaction of Intelligent Species Theory, The Future of Humanity

The Future of Humanity (cont.)

Let me begin with the observation that our "Planet Earth" is our "cradle", and our "Milky Way Galaxy" is our "Country". I hope one understands this symbolic example, using known existences to compare to real realities "out there".

It looks like a big neighborhood today, (like the "baby" (humanity) in the "Cradle" can view the "Country") like planet Earth was a few thousand years ago. The important part to understand is what a huge "Country" we exist within. I hope this reasoning will change our focus from "throw a bomb in the neighboring nation" to let us all team up and take this "Country" ("Milky Way Galaxy") under control.

We need the effort of all of us to accomplish this. We have no time to play with "pity dictators", "stupid social structures", "misleading religions" that lead us nowhere, or "our comfortable Level 2 emotions" that tell us that we are really somewhere and we really count in this huge "Galaxy". I have bad, but very real news for all of you, We (Humanity) do not even register on the real scale of the "Milky Way Galaxy" yet. We have to be very careful with our (humanity) decisions, because the way we are going "we will miss the boat". We will be extinct before we will be able to discover the wonders that are waiting for us "out there". So, here is a more detailed plan (foundation for the building) that I started in Chapter 9 and I will continue it here and now.

The beginning of the process has to be established, the "one nation", "one language" planet Earth society. This has to be based on the free will of the people of the planet. They all have to be convinced that it is in the urgent need of humanity. One of the **most important** aspects that I described in Chapter 9, is that **All Nations should keep** their native language "as a second language", keep their history, heritage, and cultural contribution to humanity through this process. We (humanity) need to merge into one society with peaceful reforms for the interest of all of us!

The population control to reduce our Earth population is the next very important factor. All the people on planet Earth has to understand the concept of family planning, and the need to reduce the population on Earth peacefully to 1 to 2.5 billion. That has to apply proportionately for every nation on the planet.

Part 2 - Chapter 16
Interaction of Intelligent Species Theory, The Future of Humanity

The Future of Humanity (cont.)

One can use the percentage calculation, for example the USA has 5% of the world's population, (300 million out of 6 billion) so, the USA should contribute 5% of the population reduction control also! I understand that it is a great sacrifice for everybody and every nation, but one needs to understand that if one really loves one's own children and grandchildren, that one has to create an environment for them in the future, where they can really live up to their best potential!

All those who "jump in" and help, taking a sacrifice, needs to know that they are really building the future for humanity. One needs to understand that population control can't be a "God's standards by the church" or "birth control law by the governments" issue. This issue is "**far bigger**" than any of those "**small minds**" that try use their power to control and play with the ever exploding population on planet Earth! Make no mistake about it, if we all turned into an "Eastern Island Scenario" (human extinction from an Island scale to a planet scale!), God won't be there to help! He didn't help them! (Where was "HE"?!). We (humanity) have to take charge, like it or not!

The implementation of the "Fair Share Distribution System" in every country can eliminate the financial problems and world hunger. The wealthy individuals of the society need to understand unless they lead the transformation to the "Fair Share Distribution System", they are actually the "enemies of humanity".

One needs to realize that one can sit in one's 24 bedroom castle with 10 super expensive cars in the garage, but if a 10 mile radius meteorite hits the planet, all that one has is irrelevant, including the one! So, my argument here is to enhance our control over our global environment (first the Solar system) so one can ensure the survival of humanity!

We (humanity) have to gain total control over diseases. (Vaccination, Medicines to stop them — even using the results of our genetic discoveries!) We need to stop creating "Chemical and Biological Weapons" using our knowledge of chemistry and human biology. One needs to understand that an accidental mutation of an "Airborne X Bug" can kill us all without the help of our human "Chemical and Biological" smart engineering process.

Part 2 - Chapter 16
Interaction of Intelligent Species Theory, The Future of Humanity

The Future of Humanity (cont.)

If one thinks it sounds idealistic, one needs to understands that all those "losers" who are playing with that deadly stuff for their advantage are in their power game, but they don't realize that they are enemies of humanity! They work against the survival of the human race, and have to be dealt with in a most serious way!

Humanity must learn from nature to make all manmade products 100% recyclable as every existing thing is in nature. Nature has a 0 pollution system. We (humanity) should copy that for our own benefit. One needs to remember that "fresh air" is a very important ("Attribute of Humanity"), and we can't exist without it! I will be very sad to find out that we were starting to care at the time when we were starting to run out of fresh air! (That would be the typical emotion based (Level 2) reactive humanity, instead of a logical based (Level 3) proactive one!)

The "New Value System" has to be developed for humanity. Something simple and easy to understand, practical, and "real freedom" driven, to guide humanity to the achievements of the control over our "universe". (More detail in Chapter 17, where I will analyze existing and a possible new "Value System").

Finally, but just as important as the others, "**the new concept**" for humanity, is that "**We have to stop being Earthbound!**". In other words, we have to start building our life and our existence to be able to **move out and exist in space**. Sounds impossible, stupid, ridiculous, maybe, but we have no other choice. Relocate, recreate our real home environment in space, and learn to live at the final frontier, or disappear from the intelligent existence.

Today in the USA we spend about 16 billion for space exploration, and about 400 billion for the armed forces. (We say for protection of our national interest (USA), but lately we pick a fight with every possible place on Earth and we have troops stationed everywhere. Why, if we are a peaceful, defensive minded people?) Maybe we are protecting the interest of the group of our wealthy people who make big bucks on the oil fields on this planet that they own, or the businesses that are located all around the world.

Part 2 - Chapter 16
Interaction of Intelligent Species Theory, The Future of Humanity

The Future of Humanity (cont.)

I hope that as a leading country that the USA one day wakes up and gives the choices to the rest of the world. **First** choice is to join us and start forming a new "One nation, One Language" society, and use all our resources together to build our future in space. **Second** choice, that one is against us, and then one is against humanity's survival, so we need to deal with that one not to make our effort impossible. **Third** choice is to stay neutral in this process, go on your own, but then we must stop all the economic or any other activities with those countries.

No food aid, no international monetary fund money, no technical aid, no military consulting, Nothing for them! These conditions sound fairly logical to me if the human race's existence is at stake! Of course, we don't need to make this a rigid arrangement, anyone who changes one's mind later is welcome to the group, as long as that one accepts the new structure and conditions of living together. Simple, working, and practical approach for the future of humanity! One needs to understand that we have to gain control over our destiny against the "Mega Tsunamis, Mega Volcanoes, and the Giant incoming Meteorites". One needs to understand that it is only a matter of time before our planet will be uninhabitable for humanity. (Ask our scientists for a change instead of believing that God will save us! He won't!)

Here is the next part of the future. Humanity has to design a "Nature Independent" existence. Starting with minimal interference from "Nature". For starters we can build our existence, like our houses, our cities, and our states in a nature secluded environment. It may sound like science fiction, but picture a dome over a city. We already have built sport stadiums that have a closed environment from weather and the outside "world". This concept can start to separate an "Artificial Human" environment from the activities and effects of nature. The next step is to use our advantage that we are the only species on this planet, (because the Level 3 Logical / Reasoning mind) that isn't Earthbound. We (humanity) are the only one who can travel outside the boundaries of planet Earth into **Space**.

Part 2 - Chapter 16
Interaction of Intelligent Species Theory, The Future of Humanity

The Future of Humanity (cont.)

We need to start building spacecraft, that are capable of orbiting around our planet, and have an artificial gravity, have a food and life support supply for about a year for 4 to 8 people, and that have all the luxuries as our current home (USA).

Sounds impossible, just like the idea of walking on the moon, but we did it. One can combine our (USA) computer science, spacecraft, airplane, submarine, RV, mobile home, and vertical lift technology — combine it with a little more USA engineering ingenuity — and there we are orbiting our home in space. The next step is to extend the size to a small town, then a city, then a state, then a country, and then the planet. Involve every country that understands the importance and supports the survival of our human race. Once we are home in space, we can orbit around other planets or stars in our own artificial man made environment.

Is it possible? I don't see any limitations on our mind other than our own lack of imagination. Here is an example for that: We try to build robots, copying our own human structure — like standing on two legs, it has arms, a head, a human like voice, etc. Very little or lack of our imagination. Think about the following idea, based on a little more free thinking, instead of copying ourselves.

Currently we have tractors, that work on the crop fields, that are built for human comfort. The cabin is equipped with air conditioning, a CD player, power stuff, etc. This is totally understandable because the operator of those vehicles works on the field in the Spring time, 12 to 16 hours a day. So, they deserve a "little comfort" on the job. Then we try to build robots that look like us, walk like us, and probably will drive a tractor one day like us. What a crazy concept! In this example: Why do we want to build a man-like robot?, to sit in the expensive air conditioned CD equipped cabin, when a "simple computer box" can be programmed for that robotic function? That can drive the tractor 24 hours on the field doing the cultivation of the land.

Saving tons of resources, because we don't have to build fancy cabins for the "man" on those tractors anymore. Simple, logical thinking. We can have the new "Fair Share Distribution System with "robots" as "slaves".

Part 2 - Chapter 16
Interaction of Intelligent Species Theory, The Future of Humanity

The Future of Humanity (cont.)

(remember robots are machines, not humans, they don't feel, or think about discrimination, they only do what we will program into them!). They can work 24 hours a day, and they won't care, because they are only machines with no emotions and with a lot better 24 hour performance than humans. If one thinks it is impossible, I don't think our technology is too far away from achieving something like that, it is more a political issue — where to spend our capital, by a vote for the "pork projects" by our congress (USA).

Time for humanity to employ all the technology that we invented in the last century for our new direction and our new survival activities in the next up-coming centuries. I am positive we (humanity) can do it.

The next important task for science is that we need to create our "water and food", but not in the traditional way. We have to invent technology that can create a desired food, without the regular "nature involved" process. Again, if one says it is an impossible mind exercise, I have to remind that one that we already (in the USA) have pre-processed food, like TV dinners, cereals, etc., where someone else, (food company) did the cooking or mixing the proper amount of needed vitamins and minerals, in our food. So the technologies are already partially out there, we (humanity) just need to define a new direction, a new concept of living and existence for them. They will do the rest, I am positive about that. One can see the validity of these arguments, that we are ready to start moving out in space.

One more observation about our current space program.

Seems to me that we are introduced everywhere in our life to a "fate based" engineering. If you look at the "USA Space Shuttle", an engineering marvel, it is one of the worst concepts that I have ever seen. One forgets, that the goal is to take control over our environment, the forces of nature and every technical tool we build. How can someone come up with the idea that something will fall back on Earth with 20-23 times the speed of sound, and glide down to landing, based on "fate", that everything will go well? And if it doesn't, we get reminded that we took a "big risk". Has anyone heard about "risk management" as a Science!? and "Not a Religion!".

Part 2 - Chapter 16
Interaction of Intelligent Species Theory, The Future of Humanity

The Future of Humanity (cont.)

How can any "human being" ask another "human being" to glide down to Earth with 23 times faster than the speed of sound, with no control over what happens!?

Has anyone ever heard about the vertical landing technology? If you can't incorporate that into flying, (like the moon landing vehicle) without some "back ups such as parachute" technology, then don't fly!

One can imagine the effort that these future plans and adventures will require from humanity. The result will be like basic army training. That level of involvement doesn't leave time for bad behavior, just good discipline. We need to understand that just like the "Self Preservation" is the highest order for the individual response, it should be the highest order for humanity to preserve itself. No one individual, dictator, king, billionaire, religion, or any group of these should be allowed to alter the path and take humanity's mind off from the number one rule — the preservation of the self. (humanity). If anybody stands to block the way, "Sorry, you must go"! We need to move out to space as fast as we can. We need to master the forces around us. Forget the waiting for God. We have to do it for ourselves. Those of you who don't care, I have news: When the time comes, we will leave and you all can stay. (The same story as Noah and his ark, that example I am sure will ring a bell!)

One needs to understand that "**The Matter and Mind complexity that creates this level of human existence is a gift of chance**" — we should not throw it away, because of our ignorance or stupidity. So far the future looks good in this chapter, looks like a working, logical plan. (one can think it over for one's own mind!) One more advancement, to move from the "Fair Share Distribution System" to the "Artificial Need-Based Free Distribution System". (In short "the production is robotics based", "the distribution is free and need based". One needs to understand that this will require different individuals, with different value systems, that may not exist today on planet Earth.)

Here is an interesting encounter that I personally experienced, and gave me "the free distribution system" idea. One place where I was consulting as a computer consultant, the management decided that they will open up a coffee vending machine for free coffee. So one didn't have to have a coin, just push the button and the coffee is there.

Part 2 - Chapter 16
Interaction of Intelligent Species Theory, The Future of Humanity

The Future of Humanity (cont.)

It was a really great experiment. In the beginning, everything was fine, people were living (in this experiment) in the "Artificial Need-Based Free Distribution System" as far as coffee distribution goes. However, the night shift arrived and that proved my criteria that one needs a different society with a different value system to understand the need-based free distribution. By the morning, the vending machine and the area was "trashed". The machine was empty, paper coffee cups were all over, some filled with coffee all the way, some half way, quarter way, etc.

The reason I know this because in the morning, I was checking that the experiment was still going and could I get one more free coffee. Seeing the environment, I realized that the animals in the wild are better than some of the low life humans. Animals use everything on the need-base and they don't "trash" their own environment. We are very far a way, unfortunately, to introduce the Free need-based distribution for the 21st century humanity. I guess before we throw away our current "slavery systems" we better educate our losers in our societies. They are not ready for the "real deal" yet. I am not sarcastic, simply concluding on my own personal observation of that experiment.

Need-based required logic (Level 3) and understanding of our environment, that has to override the emotion based (Level 2) "I want everything in life without limit" approach. I think humanity has a lot to decide for the future.

Part 2 - Chapter 17
Philosophy

Philosophy

The "Famous Chapter 17!?", (if not yet, will it be?!) that is the place, where I promised many times in this book, to provide more explanations and proofs. It is time to pick up and extend some of the thoughts from the previous chapters of Parts 1 and 2 and answer (or least try to answer) all those unanswered questions and extend on some of the thoughts that I referred to in other chapters. I didn't want to break the flow of thoughts of the previous chapters with long explanations in the middle of the point I was trying to make. Here in Chapter 17, I will follow up on all those thoughts, round up my arguments, line up my proofs, and present my philosophy of McKaneism as the complete thought system.

<u>This chapter will be divided into five main sections as follows:</u>

1.   In section one I will investigate the detailed history of human "social structures" starting from the "Tribal Structure" and ending with the "Capitalist Structure", including the analysis of the problems of the existing "World Capitalism".

2.   In section two I will take a look at the history of human philosophy, making my argument in detailed fashion that none of the previous philosophies had a comprehensive thought process as does McKaneism, in other words, by the real definition of the "science of the sciences"; none of them can stand their own ground. (**Maybe** with one exception, Aristotle!)

3.   In section three I will revisit my promise to analyze the logical thinking and reasoning mind, including the "specific", "interval", and "general" rules of our logical thinking and that reflection in our existence, that I called the "Realistic Logicism".

4.   In section four I will address the different "Value Systems" and their impact on the society, and take some time to analyze the consequence of "individual vs. society" issues. I will also introduce a new "Value System" concept with the "Fair Share Distribution System" and the theory of the "Artificial Need Based Free Distribution System", as a "First Level" and the possible "Second Level" (Chart 16) built on it.

Part 2 - Chapter 17
Philosophy

Philosophy, Section One

5.　　In section five I will revisit my arguments from Chapter 10, to place "Philosophy" in the knowledge of mankind as a "Science of the Sciences" and I will extend my Chart 26, (for your and even my own surprise, that I can get even more complicated!) so that the final full picture of "The Universe" and Philosophy is placed within it. I will also make my case for an independent "Free Institute of Philosophy" that is uncensored, and free to develop, express, and voice any opinion to guide humanity into the future.

I hope that all my readers, who took the journey this far with me will find most of their unanswered questions answered in this chapter. Let us get into a final but extensively detailed investigation of a few more and very important and exciting topics of my philosophy. Let us jump in and start with the "History of Humanity" from the view point of "social structures" of the last 100,000 years of human history as our scientists understand today, with the interpretation of McKaneism.

Section One,　　the investigation of social structures. My constant search for **"freedom"**, and the idea to change society for the better, has always been a driving force in my life. Mostly I tried my best to do it in my small humble environment. (Some success, that on one occasion I voiced my opinion at work so forcefully that actually I started my own company (that time) to change their 401K investment plan and pick a better one — they did!) Small changes, until this book. This is, I think, the big one!

I introduced a logical argument for all who are willing to think on his or her own. I am positive that the thinking process will start a change in the world for the better. (My mind is maybe coming from the late 60's early 70's when peoples were thinking that to change the world, it was for the better, and this is a "good thing".) Unfortunately, in the year 2003, many individuals I talked to believed (very wrongly) that if they change their world for the better, inside their fences and their houses, then the world would be a better place one house at a time, and "no worries" they did their duty.

Part 2 - Chapter 17
Philosophy

Philosophy, Section One (cont.)

Nobel, honest thinking, but dead wrong. We will see in this chapter the problems with the "Individuals of Society" in "Capitalism". One day we (humanity) have to wake up and realize that "we (humanity) are traveling in the same boat", (one small planet traveling in space) and it isn't a matter of who pokes a hole in the boat, we need only one hole to sink the boat. For example, one can make a most beautiful "dollhouse" out of one's home, yet it counts for very little, if non-drinkable water is coming out of the faucets or toxic chemicals fill up one's small creek in their backyard. Even if you make your home perfect, those things are coming from outside your domain. This example, even if it may be a little too strong, tries to point out that there is no such thing on planet Earth as "Private Individual Anything"! We all are bound together for better or worse. Currently in 2003, the beginning of 21st century, it is for the worse! (But that is alright, it leaves more room for this book and these thoughts!)

We are (humanity) facing the following big problems currently in the beginning of 21st century. (Just to list a few!) The planet is divided (artificially by humans) into countries and they all are searching for domination of the planet (with the use of weapons of mass destruction). The "World Capitalism" is winning and it isn't a viable long-term living solution for humanity. The "Preservation of Humanity", as an extension of the "Individual Self Preservation" (Chart 28), is overlooked by all societies. (Lack of comprehension, the higher dimensions, all this is way beyond most individuals level of thinking! — even in the leadership!) They are mostly focused on National Security, and that can destroy humanity as a whole. The different National Security issues between different countries can create a final "play off" for humanity between those nations with their powerful weapons.

The "Solutions" or a "Logical Level 3 Plan" has to come from humanity as a whole and be "Accepted" by all nations. The "Plan" has to serve humanity and not a few non-working systems or countries with their narrow minded leaders. Humanity interacts by the actions and reactions of the many individuals. Those individuals have to realize that it is possible to have a much better solution than we live by today, and we (humanity) all have to be part of the solution.

Part 2 - Chapter 17
Philosophy

Philosophy, Section One (cont.)

For making an argument of how important it is, the understanding the "Social Structures", let us investigate those that previously existed throughout human history .

One needs to understand that discovering the existing, working principals in the previous "social structures" can give us a chance to design a better one. I will start with investigating the previous "social structures" that humanity lived by, throughout human history, as we know it from our (humanity) scientists, research results, and historical writing. One needs to understand that anyone's knowledge is based on the knowledge of its predecessors. In one point of time in human history the "printing press" was invented, so we have books from that time to understand and analyze events from the past.

Before "printing", writing was established, actually thousands of years ago (in the primitive way, by cave drawing) by the early humans. We find those and carbon date those drawings in the caves, even if we can't be sure of their content all the time. It is a big "crossword puzzle" for mankind to discover it's history accurately.

In the beginning we had a **"Tribal Structure"**, the first artificial structure where each individual in that small society had his and her own chores. Males did the hunting mostly, because of their bigger physically built bodies, and females did the gathering and child raising. It was a very difficult existence. Very little protection from outside problems, like predator animals, harsh weather conditions, trouble from other tribes, etc. Every day on the planet, humanity was a throw of the dice away from extinction. Millions of years went by (about 5 million) and somehow the "Homo sapiens" appeared, around 100,000 years ago.

Here I am depending on the findings of our anthropologists, and their method of carbon dating the items they found. For a philosopher, 3 or 5 million years ago or 100,000 or 200,000 years ago, is really irrelevant. The events and their historical existence is very important, but a few years more or less in the time table doesn't change too much, as long as those events happened. Of course, I am running into the "Church" here, that believes in the creation theory, and somehow we have no mention in the Bible where those 3 to 5 million year old human-like bones are coming from. (No, it isn't Adam or Eve! Even if we named Eve the first one!) We find those remains and can carbon date them, that is a fact!

Part 2 - Chapter 17
Philosophy

Philosophy, Section One (cont.)

So, some human-like creatures were running around a few million years ago. For those who are clueless about the carbon dating techniques, an additional reading is in store as I promised in the Introduction of Part 2. Carbon dating (and they may have more modern methods that I don't know about) is very scientific and hard to argue with, unless someone wants to ignore scientific facts.

Here we are, at the beginning, and we know that "**Tribal Societies**" existed at the beginning, using as we call it "primitive" tools (primitive for us, but it was state of art for them!). Of course, nobody was there from our own timeline, so many of these thoughts are coming from observations of the remains, scientific dating techniques, and some logical reasoning based on those facts. The only direct observations we have are that we are finding small tribal groups in the Amazon rainforest, and some of those didn't get touched by our "civilization". In their isolation, they are living by their customs and beliefs, and their life is reflecting the ways that we concluded from the study of those old cave remains. They are living the same hunter-gatherer life style that we believe our ancestors did. We can't conclude 100% that they are the perfect reflection of our old times, but the similarities definitely gets one thinking.

I can make a statement here that those "**tribal societies**" were the first "**artificial social structures**" of humanity. I tried to point out that they started to alter their environment with the use of tools and the observation and use of fire. I tried before to define that those social forms that I call artificial are capable of transforming their environment for their own use. We found evidence around those old "**living quarters**" where our ancestors worked on their hunting tools, draw pictures on the wall of their hunting imaginations, or maybe hunting events. We don't find these kinds of behaviors in any known natural life form (plants or animals) on planet Earth. The most significant implication for a philosopher — that they lived and worked together in an **artificial "Tribal Structure"**, a working "**Tribal Society**". It comes as no surprise to me that the "Directional Theory" was existing in their life, as we all are affected by that, and their structure, like any other I addressed before, had a production-distribution system.

Part 2 - Chapter 17
Philosophy

Philosophy, Section One (cont.)

They were hunting and gathering until they had food for all the members of the tribe. That was the production part, and they shared the food, that was the distribution part. I am only speculating, but it seems logical that the sharing went by power structure, probably peacefully. (Something like the lion group in the safari's, where they have quarrels but no injuries within a group.)

The other very significant observation here, that they were only a few steps ahead of the "**Natural System**". It is important because in their time the exploitation of their environment didn't exist at all. These "**Tribal Structures**" lived a relative harmonious relation with their surrounding environment. They used their resources as **need-based**. That kept them in relative balance with their environment. The other interesting fact is that their harsh environment kept their number in balance. Interestingly, we don't observe big tribes with lots of individuals together in the hunter-gatherer social structure. (Like millions of people in today's countries.)

The improvement in the human mind (as one can see, a very slow early improvement, because most of the energy went for survival and food production) for the next social structure in the human history was coming alive — the "**Slavery Structure**". Some of these tribal groups became stronger than others, (having more advanced tools, more success in food production, and a greater number of individuals) and simply conquered the other tribal societies. We know very little between the time of the "Homo sapiens" 100,000 years ago and around 12,000 years ago in Egyptian culture.

We recently discovered that some of the Egyptian monuments were built during that timeline. The other issue that is still debated is that the builder of the pyramids were they "free" workers, or were they "slaves"? So our "real" but "sad" knowledge of slavery is coming from an early Greek culture.

However, we have a few things we know. Some point in time between 100,000 B.C. to 12,000 B.C. timeline, agriculture was discovered and introduced by the societies. That changed the hunter-gatherer structure, where humanity could produce their food, from 12,000 B. C. on, in a more controlled way.

Part 2 - Chapter 17
Philosophy

Philosophy, Section One (cont.)

That lead to a population explosion, and the integration of functions within those societies (we are talking here about societies in today's location of Egypt, India, and China). The food production of the new way changed the society.

Some individuals had to work in the fields doing "farming", some others had to learn about the "water and weather conditions" that farming depended on a lot, some others had to figure out the "distribution of the food" that lead to a power structure of the society, some others had to "protect the land" that lead to the establishment of the armed forces, and I can go on and on with the integration of responsibilities — that started a major separation of individual functions within the society.

The discovery that other conquered individuals can be used as a "labor force" that can free up their own individuals from the hard work of farming and provide a more pleasant life style ("Emotional Level 2 motivation!") lead to "**slavery**". One can see how this growing, evolving process of humanity ended up with a very "**Unequal Unfair**" distribution format of a "**Slavery Structure**", that is the "**Slavery Societies**". I need to point out some **very important** aspects of this structure, as I argued before, that an "**Unequal Unfair**" **production-distribution system is unsustainable**.

We discovered a very important aspect that an "Unequal Unfair" system has to be enforced on the individuals of a society. (Why do we need to enforce a sustainable existence? Logic says that if one needs to enforce something, then the one has many individuals who would not choose that system without the enforcement!) One more extremely important fact, that comes as no surprise to us by now, that those societies were formed on the 2nd Level (Emotional) human attribute.

It is easy to see from these observations that these straight forward conclusions follow: They needed an organizational structure to uphold that non-sustainable social format. They all had strong leadership, built from a group of people. They all had a **Head of the society**, (King, Ruler, Emperor, etc.) who was a symbol of their existence. They all had **strong armed forces**, to enforce the rules that their leaders created for them to live by.

Part 2 - Chapter 17
Philosophy

Philosophy, Section One (cont.)

They all had (and here comes a very important observation!) a "**Philosophy (actually in reality a thought process)**" and with this the inventors, holders, and carriers of the thoughts, their **religious leaders** (High Priests). One can easily see that no matter how strong any society is, the strongest are bound between the individuals, not the fear from the power of their King or their army, but the **belief system** that holds them together.

Worshiping a higher unknown, untouchable power, that can govern their weather, control their floods, shield them against their enemies, and constantly looking after them from some secret place. Sound familiar? One can find this "Higher Power" in the structure of most societies (because they are Emotion based!). That is the power, the glue that holds those societies together. If one is looking at it a bit closer, most "Emotionally" outbursting activities are related to their "Higher Power".

They are praying to that "Power" (some cultures had multiple "Higher Powers", some had only one!) at harvest time, thanking "it" for their good fortune, or praising "it" during the time to get married, or time to welcome their newborn, and/or time when they pass away, and "obviously" ending up in the domain of their "Higher Power". Just think back to our Knowledge Chart, (Chart 29), and it is easy to see that the individuals of these societies, who are raised and educated according their belief systems, and so by the time they become a productive member of their societies, the basic rules of existence are "carved" into them. Their "Knowledge Chart" is empty on alternative solutions. They have no knowledge of evolution, or other thoughts, or philosophies. One dimensional upbringing (looks familiar?), it was **Only** 12,000 years ago! (Surprise!: We still didn't change?!).

Works "fine" for a limited time in human history, but these systems ignored one of the strongest human emotions, and this is also at the line of the 2nd Level (Emotional) and 3rd Level (Logical/Reasoning) — it is the issue of "**Fairness!**".

No matter how hard their system tried to suppress individuals in the "**Slavery Societies**", the system's time was up. People didn't want to submit to slavery, and couldn't be held slaves forever.

Part 2 - Chapter 17
Philosophy

Philosophy, Section One (cont.)

Unfortunately, because humanity never planned their social structure before in history, (I am the first one who made a very strong argument, that our (human) systems are artificial and we should plan an "optimum"— just remember optimum is relative, we will address that later — one for our existence!) the change came from a lot of miserable wars and hardship of humanity. The "**Slavery Society**" finally ended.

It was time for the next social structure "**Feudalism**". I will investigate this structure next. We will see that we (humanity) were not much better off with this change, but the process of the "Directional Theory" kept moving on and created a better (even if it is not much better) format for human existence. I want to emphasize here the importance of "**Fairness**" as a moving, motivating factor. "**Feudalism**" was the new structure, or is it?

At the time societies transfered to Feudalism, the basic production-distribution system was already in place. The positions were already set in the society. One can find (as we saw before) the top leader (King) of the society, then the high ranking military personnel (Generals), then the land owners where the food production takes place (Landlords), and of course the very important group, the leaders of the church (High Priests) of the society. It is easy to see that individuals in these groups had privileges compared to the rest. Also, their children had a greater head start in life than the others. That is the beginning, in the "**Slavery Societies**", and continuation on in "**Feudalism**", then "**Capitalism**", and interestingly that we later called "**Socialism, Communism**", where these groups develop into "classes". That terminology was used by Marx, when he was analyzing the warfare between these groups.

McKaneism is a lot more advanced philosophy! In my observations, investigations, and conclusions, the "class" issue is unimportant. Those systems can be summarized into "Unequal Unfair" distribution systems, and the motivating factor for their collision is the "**Fairness**" issue, on the emotional and logical level, and not their "class" organization. You can call them "classes" or "groups" or any other names, but the drive for the "**Fair Share Distribution System**" is the true cause for the changes in their social structure.

Part 2 - Chapter 17
Philosophy

Philosophy, Section One (cont.)

After this little detour, let us continue with "**Feudalism**". The name came from the "**feudal**" landlords who owned the land and were a foundation of the "**Feudal Society**". Not surprising again, that the food production was in focus, because that takes place on the lands. The "Slavery Social Structure" was gone but that doesn't mean that freedom was there. At the beginning of "Feudalism" the workers of the land lived and worked on the land, but they didn't own the land and they had no rights to leave the land. This was a better deal in this society for them, better than the slavery structure, because after they paid the "**TAX**" (mostly in products!) for the King (Government, Army), then a Church, then their Landlord, then they could have whatever was left over for their products for themselves — if it was a "good year" and they had any!

The slaves in the slave societies only received as much from their production that was keeping them alive, so they could work yet another day for their holders. Some luckier workers (called peasants) in the feudal society, who had a more generous Landlord, actually had a much better and definitely "**more fair**" life than the workers of the slave societies, the slaves. After a few uprisings here and there, in Europe, the peasants won the right to be able to leave the land of their landlords if they wanted to do so (work for another landlord or go to the city!).

That was a big step, because it opened up the possibility of the industry to begin expanding. Until then, the industry in "Feudalism" was confined to bigger towns and cities, where it was mostly focused on industrial products like clothing, shoes, household items like furniture, and kitchen stuff, etc. Now some of the kids of the peasants had a chance to migrate to the towns and cities and free themselves from the farm labor for good.

We can see here that regardless of all those improvements, the "Unequal Unfair" structure remained the same. The privileged groups of people had a better life, the non-privileged ones carried the burdens of their society.

That went on about 8 to 900 years, then with the Industrial Revolution in England, "**Capitalism**" arrived.

Part 2 - Chapter 17
Philosophy

Philosophy, Section One (cont.)

Still interesting to notice, that is was unplanned, and it only happened by chance as the society transformed into a different living existence by the improvement of it's technology. "**Capitalism**" was the new structure of the society. With the beginning of the Industrial Revolution the production-distribution system had to change from Feudalism to Capitalism. We saw this earlier, then from the viewpoint of the philosophy McKaneism, Capitalism is an "Unequal Unfair" distribution system, just like all the previous ones. The most powerful feature of Capitalism, however, is the introduction of "**Capital**", where the system gets it's name from. The capital is an existing "fund" most of the time in money format, but it can be gold, or any other form of valued items of the society; like another example diamonds or even legal obligations. These "Funds", or we can call it "Wealth", can be transformed into working capital. The most important features of the working capital are that they finance some production ventures, and gain an extra return on the finance.

For example: a capital firm finances the production of bread, and after the sale to the consumers, we end up with more capital than we started with. Anyone can see that there is a **risk** involved in this process, so if we calculate our risk poorly, we can end up with less capital than we started with. That nature of the capital creates incredible possibilities for advancements and also for failures.

That started the restructuring of the societies from Feudalism to Capitalism. We still have to notice here that the leaders of the feudal society (Kings, High ranking Government and Military people and Priests and the Landlords) were in the "driver's seat" because of the level of wealth they had to start this transformation process. So the "Unequal Unfair" nature of the society remained. One of the greatest books coming out was from Marx, "Capital" in German language and in many other language translation. Of course, in the Capitalist world, his book was on the black list and he himself considered an evil person, because he concluded, after a long period of research and investigation (about 20 years), that Capitalism must be replaced with Communism.

I was born and grew up in communism, and it is very interesting how many people, even in the communist countries, have no clue about what Communism really is.

Part 2 - Chapter 17
Philosophy

Philosophy, Section One (cont.)

I will return to this issue later, because after Capitalism in 1917 in the Soviet Union, communism actually became alive. Before we get there we need to investigate one very important aspect of capitalism. The basic formula "Capital"-"Something Happens"-"More Capital". That is the formula of capitalism in a "nut shell". Interestingly, Marx tried to investigate these events, and he found himself in the world of economy, and opened himself up for a lot of criticism from leading economists. I don't make this same mistake, because philosophy shouldn't try to explain the economy or economic principles. Philosophy is here to understand, organize, and induce other sciences, and to have a healthy action - reaction relation with them both ways.

Let us investigate the "Capital"-"Something Happens"-"More Capital" formula. The first part is simple, I have the "Capital" in my pocket. The interesting part comes from the next step.

The question is, if I have 1,000 in Capital, how does this amount turn into 2,000? No surprise, there isn't any magic involved here at all. We know that the 1,000 will go into a production-distribution system to fund some functions. Let us look at some very simple examples for analyzing this process from the philosophical point of view.

Let us look into a pizza venture, to make one pizza. I need the basic materials, the heat of the oven, and of course the employee who does the job. (I know there is a lot more involved here, but this simple example will be proper to make my points.) Our goal is to sell the pizza and make more than we started with after we've paid for everything else. It is easy to see in this example that one can't make a huge profit on the basic materials, so **the profit has to come from paying the employee less than his "contribution" or "labor" is worth**. Think about it, how many pizzas can be made in one hour by that employee who makes 7 or 8 bucks an hour?

That is the basic concept of Capitalism. People who were advocating communism called this the "exploitation of labor". Of course, that made them evil in the eyes of the "religious" capitalists. One needs to understand that without profit or getting back "More Capital", there won't be any incentive to run any production ventures.

Part 2 - Chapter 17
Philosophy

Philosophy, Section One (cont.)

So, this formula so far is OK! The problem begins when one can own a "Chain of Pizza Stores", and can have a huge amount of profit, placing that person in the position to control and influence society with his wealth like the Landlords did in Feudalism. They can make decisions how to invest their money into different ventures. Holding large capital in capitalism (Billions) gives one a larger power over a lot of things in that system. I will revisit this issue later when we arrive at the concept of the "Fair Share Distribution System".

For now, we have to take the next historical step — 1917, that is the formation of the communist society. "**Communism**" is a social structure that we've heard a lot about all over the world. But very few people really understand the concept, philosophy, and its actual realization, not even most of those people in the societies that lived or currently live in that structure.

The communist concept is an "Equal Fair" distribution system. This is great in concept but **it can't work in reality**. Let us see why. I don't have to tell a lot about it if I distribute something "Equally", then my distribution is automatically "Fair". For example, take five people, each gets the same size and number of (3) apples. (equal distribution) No room for any complaints about equality and fairness. Looks good and feels good, so what is the problem?

The problem is that this system doesn't work! (I have to say here that the western world thought that they beat communism in the "cold war". Nice to think that, but it isn't the truth.) The 1917 Russian revolution changed their society into a communist society. Here is a few problems that very few people know or understand. If you read Marx about communism, he said that communism will win first in the country that reached the highest level of capitalism, (final stage of Capitalism) and he called it Imperialism.

He argued that in Imperialism, the society has plenty of everything to distribute equally. One needs a very advanced and very rich society with an abundance of production and products. Think about it, even in the USA, we don't have that level of wealth. For example, that means **everybody** can get a seventy foot sailboat "**equally**" (about 3 million bucks each!).

217

Part 2 - Chapter 17
Philosophy

Philosophy, Section One (cont.)

One can see the criteria of equal distribution is that whatever products we have, we have to have "it" for every individual in the society! No society ever achieved that level of wealth on planet Earth yet! That makes you think: What system were those countries living in that we called Socialist or Communist? They were not communist by the definition of Marx!

To complicate matters for the worse, here comes Lenin, who's statement was that: "The chain is always broken at the weakest point!" (loose translation from me). He tried to imply that the Russian empire was the weakest capitalist country (was at that time on Earth) so it was the easiest to take down. I hope my readers start to understand this conflict between the two schools of thought. They both were agreeing on the "Equal Fair" share distribution system, but they (Marx, Lenin) were coming from a different direction, ending up with different conclusions.

So what is it that really happened in reality?

In reality, the communist change targeted the leadership (Class Warfare) of the country, believing that they were the suppressors of the people and their freedom. They went against the "Czar, the High Priests, Landlords, and of course the Capitalists who hold the capital." The concept was to eliminate those positions and change the social structure. They killed the royal family, during the Russian revolution, and the same thing happened with those Landlords and Capitalists who didn't leave the country in time.

That was a bloody exchange in the society. (Revolution, a simple to complex fast Directional Change in the society, actually a fast complex to a more complex!) To eliminate the Religion, they ran into some trouble. Religion runs deep in the people's emotional world, so one can't turn that off like a light switch.

They came up with an atheist philosophy, that denounced the existence of God, and started to teach it in the school system from the age of six for the new generation. Around 1919, the process was completed! Or was it ?!! Let us look closer at what they did, and why it didn't work for them in the long run.

Part 2 - Chapter 17
Philosophy

Philosophy, Section One (cont.)

Problem number one, they didn't have a society producing plentiful goods and services that they can share equally. As Marx said "it has to be an attribute of communism". He was correct in his argument against Lenin (the weak chain concept is wrong!). So, we have a society between "Feudalism and Capitalism" (Russia in 1917!) and they thought they could cruise into "Communism" with the existence of plenty! That was wishful thinking, but not reality.

Problem number two, if one takes away the incentive of the capitalist to make profits, then they also take away the incentive to be creative, to work harder than others, to stand out with their products and services. In other words, they ended up going from one day to the other with the lack of incentives and knowledge to run the country.

Problem number three, they replaced the "**old** leading class" with a "**new** leading class" from the poor people by force, who became just as spoiled and corrupt as the old rich leading class was. So the old game didn't change, only new players were sitting at the table! One thing they enforced, based on the communist teaching, was that no one could own any capital and make profits on others, so the state owned everything. Of course, they ended up with a State run structure, where nobody (individually) owns anything and nobody (individually) was responsible for anything (A recipe for disaster!).

These three problems are only the major problems among many other small ones. One needs to study history to get the whole picture. I lived in that structure for more than 25 years, and to detail all the problems, it could take another book much larger in volume than this one. I concluded many times in my analysis, that from the viewpoint of philosophy, those major problems prove my point of a non-working social structure. The "new game was in town", that if one was in the leadership and in the communist party (that was a given for success) that one can have anything one wants, but with one condition, that one has to keep one's privileges quiet. For a sad example: A 10 million dollar estate for an East German leader was discovered at the falling of the Berlin wall in 1990-92, with a 10,000 bottle winery in the basement worth a few million — in the communist society where all should have an "Equal Fair" distribution system by the "Book of Communism!".

Part 2 - Chapter 17
Philosophy

Philosophy, Section One (cont.)

Like I said before, the old bandits were replaced with the new bandits. So, communism ended because of their failure in most countries. The few dictators with their bandits who still are in power today in some countries (as of 2003), will disappear as soon as the people discover a better system. (McKaneist "Unequal, Fair Share Distribution System" will do the trick for them if they can understand and comprehend it!)

One more interesting and "personal" comment about communism. It was hard for me to see, that people "dragged down the statue of Lenin" (No surprise in a sense, a fellow Philosopher!) — maybe some of the same people that put it there in the first place! Lenin was a great philosopher and a dreamer. Lenin was actually one of those who discovered that society needs to be analyzed for it's structure and that people have to work on to come up with the best possible "Social Structure".

Lenin was living in the time of the Russian empire when the Russian people were very suppressed, and most of them had a miserable living condition. He spent all of his life building his philosophy and lead the people out of their misery. I have had a chance to read all his books (but 3) in Hungarian translation. I can highly recommend the "State and the Revolution" to anyone who wants to read very logical, well constructed arguments about the class structure of societies (It will make you think about the statement "We the People!"). Most of his publications were teaching the people and instructing them on how to go about the abolishment of the old and the establishment of the new country.

I am still wondering, what was the content in those 3 books that I didn't even find in the Hungarian communist library?! His unfortunate, quick but quiet departure, at the time when Stalin the Dictator rose to power, is still a mystery in my (and many others) mind. (A long lasting tradition was alive in the Castles of Russia's Czars, that many people had died quietly by poison! It would be interesting to study the sudden death of Lenin, who was a great leader and a dreamer, but not a brutal dictator such as Stalin, who followed him in power! One more sad example that "Philosophers" were not and have not been well received in their societies most of the time!)

Part 2 - Chapter 17
Philosophy

Philosophy, Section One (cont.)

Now, one can see "how", in the so called communist societies, some roothless opportunists take over the society ("high jacked") and rule it, living better than the Capitalists or the King themselves, of course very secretively. Having it all in the inner circles! These were the main reasons why "**Communism**" collapsed in the Eastern block, **they went bankrupt with their non-working, communist production-distribution "social structure".**

The correct logical conclusion is that the competition with the West helped the communist system collapse **faster, ahead of time**, but the reason was their non-working social system, not the victory of the "cold war"! The world is back to world wide "Capitalism", and everything is fine.

Not exactly!

"**Capitalism**" is a non-working system, just as the others before it. Of course, non-working is a relative term in the scale of a historical time table. A few hundred years in history is nothing compared to the millions or billions of years of the "Universe". So, when I am making a statement that capitalism doesn't work, I mean that this system will fade away like any of the others, and will leave room for a new, better, and more suitable system for humanity.

Before I move on, I have to analyze a few more very important points here! One can see from history that the transformation from one "social structures" to the next is a gradual process. In history, these events take place involving many years, decades, or even centuries. One can't go to bed at night in "Feudalism" and wake up in the morning in "Capitalism". Sometimes it is difficult to see the basic structure because these social systems overlap in the process of transformation, but under closer investigation can reveal the "Unequal Unfair" nature of all of these previous systems.

The next important observation, is that we know from the "Roman Empire" (maybe earlier) that they introduced the concepts of "Taxation". The conquering power didn't physically remove everybody from the conquered country.

Part 2 - Chapter 17
Philosophy

Philosophy, Section One (cont.)

They took a large number of "slaves" to work for them in their own country (as slave butlers, maids, personal teachers for the rich kids, workers for the farms and road construction, gladiators for their "Brutal Entertainment!" and on and on), but they "**enslaved**" the rest of the conquered country **through** "**Taxation**"! They created a position of "Ruler of the Colony", and that person collected the most aggressive taxes from everybody that lived there. The fear of retribution from the "Roman Army" kept the colony obedient! It is important to point out that "Taxation", a "Logical Level 3 tool", helped the ruling groups control the rest! Another example that even if the society is emotionally driven, the logical level of thinking is also involved. Logic isn't involved in the best design of the society, but it is involved to control the masses in the society! Changing to "Feudalism", then "Capitalism", one can see that this system of "Taxation" can keep the control over the population of the society. In our 21st century USA, the system of Taxation is used to "socially engineer" the society and control behavior. More about this later.

I analyzed the history of the major production-distribution systems. One can see that they all get removed through history and that none of them provided a "**Fair**" living environment for all the people that lived under those societies. **Here I will introduce an** "**Extremely Important and very Powerful**" **argument — the** "**Exploitation of Attributes of Humanity**" **in every previously existed society, including** "**Capitalism**"! We know this from the Attributes of Humanity in Chapter 5 (Attributes of Human Beings Theory!) and in more detail in Chapter 13 (The extended Attributes of Humanity!). One can see from those arguments that we (Human beings) are living within our "Natural Boundaries"! We keep extending them, but we can't change them. I also argued that these attributes define ourselves as Human beings.

Now it is time to analyze how these attributes get exploited in the human societies by one group of people to gain control over the other group of people. One needs to understand that the "**Emotional status to Feel Good**" was an important drive throughout human history. Most of the time in history, the leaders of the societies **justified for themselves** that they were doing a service to the rest.

Part 2 - Chapter 17
Philosophy

Philosophy, Section One (cont.)

Leading them (Kings), showing great peace for them through religions (High Priests), providing them great hopes in life for wealth (Capitalists), giving them a great chance to work, live on, or maybe own the land (Feudal Landlords). However, the underlying concept to those justifications is the motive to "**Control and Benefit**" from the others hard work. That also means that they didn't have to be part of the production-distribution system in the production side, only the distribution side. (With a lot of goodies! coming to their direction for "free", because they had privileged status in the society!)

All these privileges and power were and are held through the exploitation of the Attributes of Human Beings!

Just to remember, the basic most important Attributes: Air, Shelter, Water, Food! These are the attributes that are "Needed" for human existence. So, the leading group in each system found their own way to control the "Distribution" of those needs, then they could control their society.

For example one: in the "**Slavery Society**", if one slave refused to obey the orders, the slave holders could stop giving that one "Water and Food", so the individual can act based on the "Self Preservation" (Chart 28) or disappear from existence!

For example two: in "**Feudalism**", if one started to create disturbing thought processes, (like the Sun is center of the universe), then that one was submitted to the brutal way of "Inquisition", where in the torture chambers, they put that one's head under water so "No Air", or let them sleep on cold concrete so "No Shelter" protection from the weather, or "No Water and Food", etc.

"**Capitalism**" doesn't stay far behind either. In the "tricky way", so people don't realize the underlying truth, everything transferred to the language of money. So one buys or rents one's "Shelter (housing, clothing), Water (pay for the utility service) and Food", and for exchange, go to work 8 to 10 hours a day to have the money to continue paying for those things. And of course, "Taxation" is in order for all also! So far, it seems like humanity didn't live in a structure that didn't lead to one group exploiting the other group for the "Feels Good Emotions"! (Remember, the leading group really has it all!)

Part 2 - Chapter 17
Philosophy

Philosophy, Section One (cont.)

One more observation about "Capitalism". In the most modern societies, people can buy those basic attributes of survival on "credit". That introduces the "Liability = Slavery" society. If one goes for "Loans = Liability", that one also has to realize that those "Liabilities = Obligations"; this affects one's freedom in the society very negatively.

I hear your argument that one can control one's desires, so don't use the credit system, then that one doesn't get exploited! That is a totally loser argument. One needs to realize that one can't exist anywhere on this planet without satisfying the basic "Attributes of Human Existence"! (By the way, if one in "Capitalism" makes too much noise, like peaceful marches for "Human Rights", that one can end up with a bullet in the head!)

The message: Maybe it is not too smart going against the leading group! So far, no change from the beginning for mankind!

For example three: If one wakes up every morning, gets oneself together in 1 hour, sits between 1/2 to 2 hours in the rush hour traffic, works 8 to 12 hours, back to 1/2 to 2 hours of rush hour traffic, gets home, gets to spend about 4 to 6 hours private time, sleeps 7 to 8 hours, and then starts the cycle all over throughout the whole year — does that one "feel like a free one" or think to do this as a "freedom of choice" or as "a free individual"?

The only thing I can say: "Good Luck!, if this is the "World" one wants to live in!" That one didn't even get to the level of self-realization to understand that one's own life. That one didn't take the first "baby step" in one's human existence! Remember, that one can "Justify" the "Why?" in every situation, because the whole setting is based on "Emotions"! and the exploitation of the "Attributes of Human Beings". (One can justify, in one's life, that one does all those things for a "young new born child", or an elderly parent, or that one is taking care of another, or the pleasure and satisfaction of the weekend mind, or many other reasons!)

The point is here that "Capitalism" is just an exploiting system for the "Attributes of Human Beings" as any of the previous one's were!

Part 2 - Chapter 17
Philosophy

Philosophy, Section One (cont.)

I will investigate next the **five facts** that I think signal the non-working nature of the "Unequal, Unfair, Capitalist Distribution System" and reflect some statistics in our modern 21st century, including some data on the "State of the Planet" — and some are on the "State of the USA" as a leader of this planet. I will introduce some statistical facts here to set up the "Snapshot of Earth" and "Snapshot of the USA" in 2003.

Anyone can challenge these results of my research, however, one needs to remember that a "snapshot" only describes "one point of time" in the dynamic ever-changing scale of our societies. One also will realize through one's own research that some of this data actually gets worse from the time I take my "Snapshot". My argument here is that a "World Capitalism" is a non-working "social structure". I also will introduce the "Fair Share Distribution System" in **Section Four**, to show the contrast between a "non-working vs. a possible working" social structure. Anyone can challenge my facts, but keep it in mind, that the facts are usually not open for discussion! This is why they are called facts!

I made my statement that **"Capitalism"** is a non-working system, and thus I will investigate the following areas with the facts to prove my argument in the year 2003:

<u>Fact 1</u>. The "Closed System" nature of our planet Earth will lead to a Self-Realization that "Not Enough Wealth" is available to go around and "Capitalism" will have no answer for it!

<u>Fact 2</u>. The current "Unequal Unfair Ownership" of the natural resources on Earth will create an on going conflict between nations and interest groups!

<u>Fact 3</u>. The "Faster Population Growth" will be faster than the growth of the possible "Production" that can support the "Distribution" system in the ever growing population!

<u>Fact 4</u>. The "Exhaustion" of the "Natural Resources"!

<u>Fact 5</u>. The "Exploitation" of the "Attributes of Human Beings" leads to the realization of the lack of "Freedom" of humanity and the strong "Emotional, Logical" drive for establishing a system based on the "Fairness" and "Freedom" of the society!

Part 2 - Chapter 17
Philosophy

Philosophy, Section One (cont.)

<u>**Fact 1**</u>. Capitalism is based on the 2nd Level of mind, that is an emotional set of mind. I will take the USA for my example, but before I do that I have to repeat, as one will see after the investigation of these five major facts, that 80 to 85% of the countries on Earth today aren't worth living in! Unfortunately, their system and their people are irrelevant in the future of humanity, unless the leading 15 to 20% of the countries that are worth living in bring them into the "game plan"! The USA and a couple dozen other countries are the only ones that maybe are worth living on this planet.

I know that is a sad statistic, but we have countries where the majority of the population doesn't have a "cultured shelter", "clean water" or a "proper amount of food", (remember the Attributes of Human Beings?!) or as we say in the USA "they are under the rock!". That 80% comes close to 5 billion people out of 6 billion. That is a very bad statistic by any means.

Thanks to Capitalism, or any other non-working "Unequal Unfair" distribution system, one can be born in a country on this planet where one won't have any good opportunities for a good life at all. We call them third world, underprivileged, underdeveloped countries. How would you like your country called like that ?

"**Capitalism**" tries to base itself on the concept that anybody can have it all! That is a totally wrong concept! (Of course it "feels good emotionally"!, being an "Average Joe" in the society and thinking every day that "**they told me**" that one day I can have it all too!) Let us take some reality in the picture, that capitalism tries to hide — that "**Capitalism**" is built on the previous system of "**Feudalism**", that is it brings over the privileged groups.

In capitalism, most of the wealth is "Old Money" coming from the past (remember some countries the "Kings and Queens" still have their castles and privileges!). We also have previous generations who were there before. Especially in the USA, one can't compete with the people whose great grandfather found the oil, or the gold, or started the railroad, or the banking system, or the stock and commodity market, etc. There are a few "New Money" people out there too, but the "social structure" of Capitalism makes it extremely difficult to break into the "privileged, already wealthy class or group".

Part 2 - Chapter 17
Philosophy

Philosophy, Section One (cont.)

Here is your "anybody can have it all" (Chart 38) <u>**Individual**</u> break down: for example, in the USA (2003), the country that produces 1/4 of the wealth of planet Earth (for 5% of the population on the planet with 25% of the wealth production we can call it the richest country on the planet):

<u>**Chart 38.**</u>

1. 400 people have $880 billion wealth.
2. About 300,000 had a net worth of 10 million, about 0.23-0.25% of all the households (all my percentage (%) calculations were based on the 120-130 million household taxpayer number!).
3. About 4.6 million households have $1 million wealth, about 3.5-3.8% of all the households.
4. In the meantime (2003): about 40 million people have no health insurance (in other words, in the most advanced country, our USA, their "**good health is unimportant**"!).
5. There is more (Census Bureau 2000): about 31.1 million are considered poor, that live under the poverty level (Family of four with the annual income of $17,603 or less! And this is usually two working parents and two kids!).
6. About half of the households, 60-65 million households, had income below $42,100.

Here is your "anybody can have it all" (Chart 39) <u>**Business**</u> break down: (business distribution picture as of years 1998-2000, before the "business world" really declined "big time" in the USA and the stock markets declined by 10 *Trillion* dollars!)

<u>**Chart 39.**</u>

1. **78%** of the businesses are **Proprietorships.**
    96% (out of 78%)   sales < $100,000
    0.16% (out of 78%, about 18,000)   sales > $1,000,000

2. **8%** of the businesses are **Partnerships.**
    76% (out of 8%)   sales < $100,000
    1.8%   sales > $1,000,000

3. **14%** of the businesses are (about 2,000,000) **Corporations.**
    44% (out of 14%)   sales < $100,000
    43% (out of 14%) $100,000 < sales < $1,000,000
    13% (out of 14%)   sales > $1,000,000

Less that 1% (out of 14%), that is less than 200,000 of the Corporations, are publicly traded on the stock markets.

Part 2 - Chapter 17
Philosophy

Philosophy, Section One (cont.)

The **Capitalist** concept that "anybody can have it all!", came from the historical view of the founding fathers of the USA, because in their time (about 230 years ago) the resources outweighed the needs of the population in this new country. The opportunities seemed unlimited and have been unlimited in reality. Now, when we have 300 million people in the USA, yet 6 billion people on the planet for our resources, we (humanity) are starting to realize that we are in a "closed world", we have a "closed system", and the opportunities that "anybody" can have a "Billion Dollar" wealth is disappearing fast! The facts are that "**World Capitalism**", as a leading social system, **managed to make humanity "Broke!"** as of the year 2003.

Here is the proof. Let us take a moment for an **imaginary** "Equal Fair" distribution system and split the wealth of the world equally. According to the statistics, we (the planet) have a Planet Gross Production worth 40 trillion dollars. That is a lot of money, no question about it. How about we do the math and divide the money for 6 billion people. (that is the current statistical population of our planet.) so : $40,000,000,000,000 / 6,000,000,000 = $6,667 (with rounding). So if I want to be "**deadly fair**", like an "**Equal Fair**" distribution and divide the wealth of the planet equally, we all get $6,667 for a year (I want to see our distinguished senators and congressmen living on that kind of money, or some of our billionaires). If one remembers, the poverty level of the USA is around $17,603 for families (with two incomes), so with our hypothetical distribution, two times $6,667 is $13,334, and that puts a two income family under the poverty level in the USA, even with an equal absolutely fair distribution system applied to the planet Earth.

So, with the best, most fair intention of our hypothetical distribution, our planet Earth and it's population in the year 2003 is flat broke !!!

One can ask why on Earth would we want to divide the bucks equally? One needs to remember that, even if this is a **hypothetical example**, to bring attention to the ratio of the "Wealth of the planet vs. the planet's Population number" — the natural rule of "Equilibrium" will drive the direction of distribution to "Fairness", moving the planet's wealth in that direction, to be divided "Fairly"!

Part 2 - Chapter 17
Philosophy

Philosophy, Section One (cont.)

To conclude these observations, one can realize that our planet is a "Closed System", and with the Capitalist "Unequal Unfair" distribution, the "Unfairness" will be challenged constantly, within the country of the USA and also worldwide, internationally.

If a **few** get a hold of the bulk of the wealth, that leaves "Not Enough Wealth" to go around to the **many**. In the year 2003, we already see signs of a very "boiling" status in many countries, and make no mistake about it, it is impossible to calm down or suppress all the societies on the planet with armed forces.

A new distribution system has to be introduced for the benefit of humanity. (Anyone who wants to challenge my statistics, go ahead and do it. That one will find out that with the population growth, the ratio in the numbers gets worse in upcoming years!)

<u>Fact 2.</u> The next major problem is the ownership of the natural resources. Can one imagine that one country, because of their ancestry and location, can own 25% or more of the resources of the planet (like oil)? That basically (that will be my next argument) **belongs to <u>all the people</u> of this planet**.

I had a statement before that we are all traveling in space "in one boat", that is planet Earth! Think about it, if I say "I am the Captain" (like the privileged ones in those countries!) in our "boat", while we are sailing I own all the "Drinking Water and the Food", and I decide who gets what and when! One can wonder how many seconds will go by before mutiny explodes in my sailboat.

The same concept exists on this planet, and so far all the people are blind folded and going with it. Some regions have all the oil resources, yet that belongs to all of us, and they think they should have all the wealth and profits out of this ownership. (Humanity is very close to be a joke, in the year 2003, with their non-working broke Capitalism.)

Good luck for all the other (other than the USA) developed countries too, if you believe you can lock out the rest of the world with your army or border patrols.

Part 2 - Chapter 17
Philosophy

Philosophy, Section One (cont.)

The same truth exists for all these "Kings and Billionaires", because their ancestors created the wealth in a non-working society and handed it down to them, they think they can sit back and play the silly games of shopping for more millions, or traveling around the world carefree, or buy companies and throw everybody who use to work there out, etc.

Your social responsibility, because of that privileged wealth, should be to investigate, support, and introduce changes and reforms in the societies, to insure the well-being of the human race. Have no fear, I will do it for you! One warning, of all of you that is focused on your "one hundred wives" in your "oil kingdom", instead of the change in the human distribution system, you will disappear with the misfortunes of your humanity around you!

The sad picture is that it isn't just our oil wealth. For example, another country is planning to cut down a "tropical rain forest" covering 15% of Africa's third biggest country (that 15% is ten times the size of Switzerland). One needs to remember the "Attributes of Human Beings", that **fresh air** from our rainforests is one of the "**#1**" attributes for our survival!

I hope I made my strong point, that all the resources of the planet belong to humanity. One needs to realize that the "Fair Share Distribution System" will help one with the wealth to put humanity on the right track, and because the "Fair Share Distribution System" supports the "Unequal Fair" distribution, one can see that one doesn't need to fear the disappearance of one's wealth, in a comfortable level. There is a huge difference between "comfortable vs. unnecessarily lavish and wasteful"! One planet, one "boat", one set of resources!

Only one possible living environment. Why should one have a billion dollars and the other one should have zero? Humanity has to understand this argument if we all plan to survive as an "Intelligent race"!

Fact 3. "Faster Population Growth", faster than the growth of the possible "Production". This is just the beginning.

Part 2 - Chapter 17
Philosophy

Philosophy, Section One (cont.)

One needs to understand that many people on this planet don't reach the level of philosophical thinking, so they can't define their argument, they just know that something is wrong in here. The problem is the false sense of "freedom" in the "World Capitalism". Our views are distracted from very important problems. In our continuous party atmosphere (living by one day at a time, all the fun for me, emotional society) we forget to focus on the real problems. Think for a moment for the "Directional Theory, and the rules of Equilibrium", that all 6 billion people are coming for a party. Imagine your crowded highways now, how about when the Earth's population skyrockets to 10.9 billion, somewhere around 2050?

All this information can be found in realistic surveys. The world's 49 least developed countries will come close to tripling their populations from 668 million to 1.86 billion. They already fall into the category that I called "not worth living there". How about if they all want to migrate to the first 10 to 15 industrialized countries? What will we do? Shoot all 1.86 billion of them at the border? How about the population of 8 billion by the year 2025? One can argue, so what, our technology of agriculture will take care of all the problems.

Here is the argument for that "smart one!". Here comes McKaneism, and we will go back to the science of biology. It is good to have a philosophy that covers all the knowledge! Easy to make powerful, sensible arguments!

We know from **fact** that a crocodile needs six times it's body weight in food for a year to survive (that probably contributed to their 200 million year old existence!), the lions need 40 times their body weight in a year to survive. I don't have any data on Humans, but it is easy to see that it is measurable, the same way as for the animals!

One can ask, why is it important?

Here comes a surprise for that one! In my old research data (it is probably worse now, in the year 2003) in the USA, we had 73 individuals per square mile of land. In the world we had 110 individuals per square mile of land.

Part 2 - Chapter 17
Philosophy

Philosophy, Section One (cont.)

One can easily see that the 1 individual per square mile of land mass will determine the survival capability of humanity, just as the territory of the lion group determines their survival by their hunting area in a square mile.

The intense population growth will lead us to our next problem.

Fact 4. The overpopulating masses will have a bigger and bigger appetite for consumption, based on the "Emotional" set up of Capitalism, that could exhaust the natural resources of our planet. I already argued in Chapter 6, that if the other nations want the same amount of cars per household as in the USA, we won't (or maybe can't) support the gasoline needs for humanity.

One needs to understand that some of our natural energy resources (like oil) take hundreds of millions of years to develop in the natural environment. Once we used it all up, we are out of luck! The same can be said for all the other resources. More people need more "Shelters" (Attributes of Human Beings!), so we clear all our forests.

In my last research I found that we (humanity) have about 57 million square miles of land to live on - 33 million square miles of fertile land, 19 million square mile of steep, and 5 million square miles of deserts. One can see why I am talking about such a "Closed System" as the "Planet Earth". That is all the land we have, no less and no more. It is finite, not infinite. That is all the land we have to manage, to cultivate, to survive on whether one likes it or not!

Now, one can put two and two together, and figure out how much land is needed to produce the crops and live stock for us (humanity) to feed the "x times" our body weight per year to survive!

A very logical fact and observation based argument (like always in McKaneism), to conclude that we need to address our overpopulation if humanity wants to survive on this planet! One needs to see that the "anybody can have it all!", emotional, "Good Old Capitalism" has no answer for you!

Part 2 - Chapter 17
Philosophy

Philosophy, Section One (cont.)

Those are major problems and unless we get a lift to another planet by aliens (and we didn't sit down and talk with any volunteers for that yet!) we will exhaust our possible existence with our appetite for consumption. "Capitalism" falls apart in the future because Capitalism, as a concept, is based on unfairness and the supposed unlimited, inexhaustible resources and opportunities.

In our current world in 2003, we already have nations that can't stand on their own feet. The wealthiest countries are already carrying some of the troubled ones, but with the overpopulation affect, even the wealthiest countries won't be able to carry the rest of the world. For example, in the USA, many of us are already working 8, 10, 12, hours a day (the most productive work force, Thanks!) and part of our "Tax Dollars" go to support the needy nations.

When the population explodes to 10 billion, one may work and contribute 14, 16 hours in one's life and that may not even work! Many nations and their radical groups are already starting to use the cover of religions to attack the wealthiest countries, **not knowing that their problem is not in the religious difference, but that their basic trouble is by the unfairness of the distribution system of capitalism**. We need to wake up quick, and think about the changes, because these attacks are only the "first whistle of the pressure cooker!", it can explode, or it can get far bigger as time goes on with no changes!

<u>Fact 5.</u> I investigated in the history of "social structures" the "Exploitation" of the "Attributes of Human Beings", so I don't want to repeat myself, I just want to add a few new thoughts. I will extend on the "Freedom" issue, that we hear about hundreds of times every day! We (humanity) don't have the "freedom" as we are told many times in our life in "Capitalism". One hears about "freedom" a "million" times in one's life because that is the word that lifts one's "emotional spirit" no matter what is going on around one's life.

The "**Freedom**" that gets referred to in this society so many times is only "emotional freedom", but not "real freedom". You are free to choose, in many aspects of life, to satisfy your emotional, 2nd Level feelings.

Part 2 - Chapter 17
Philosophy

Philosophy, Section One (cont.)

You can go to the Mall and pick between the blue or red colored stuff. Once you make your decision, you feel good about it because you exercised your freedom to choose. In reality, in this society, you are just as "enslaved", like in any others before this.

Capitalism lets you accelerate your emotional feelings (even inducing them with marketing tricks) so the individual mortgage is the way to go for the next 30-40 years of his or hers earnings power. One can buy his or hers favorite house, car, boat, RV, etc. and has nothing else to do, except just go into the workplace and be a "slave" there for the next 30-40 years to pay in full all those things that the emotional state of mind of that person desired.

Make no mistake about it, the "rich dudes" didn't do too much better either. They are the ones who play their goofy games of control over the societies, competing with each other, and try to keep the rest of the people under control. That is their full time job ("their freedom"!).

However, try to tell this society that you want to stay home, read and research philosophy, and you want that society to provide you shelter, water, and food — you will find out that you are out of your luck! The "Capitalist" society is just like the others before it, and it doesn't value philosophy, because this science doesn't have an instant "cash" producing aspect like "engineering" (One has a blue print of a house, it can be sold for "cash").

And more... The leading groups in "Capitalism" are always afraid of the masses. They have a lot to lose. They are use to "good old, well established tricks" (coming from the Roman empire) to keep the masses under control! One good trick to keep the people divided is the different religious groups, different political parties, different charity organizations, etc. The next trick is the constant source of entertainment –cable TV's with hundreds of channels! (Gladiators and the Olympics in the old Roman Empire!) Then the "Free Elections", when one can believe that one is actually running his own country! (Good trick, but the people are not stupid! The USA has 120-130 million households, yet only 50 some million votes! It does not work!)

Part 2 - Chapter 17
Philosophy

Philosophy, Section One (cont.)

Aristotle argued about 2500 years ago that democracy doesn't work, and every big mouth politician in our country "salutes democracy" five times a day. Our founding fathers (USA) had a wish to make this country a "Republic". I am wondering if our leaders forgot history here?

Here is a real story for you on how the "free election?" was conducted in the Socialist-Communist country of Hungary in the mid 1970's. It has a huge lesson to learn from it. The communist party (at that time in 1970) nominated the candidates for the open positions (positions like Hungarian congressman). The "best" part of the deal was that all candidates were chosen from the "communist party members". (No membership, no candidacy!). So one could vote "freely!!!" for "Joe the communist" or "Jill the communist", who had the same affiliation, the same philosophy, and the same zero interest in the well being of the society! It was a joke! Communist democracy at work!

I have a strange feeling when I see our two party system, in the USA, where all the candidates are coming from the two parties, exclusively, locking out any other party from any debate, like they are the only important and correct ones. It is also disturbing for me that the supporting "Billionaires and Millionaires" send a contribution check, just in case, for both party candidates. (its like betting on all horses in a race, so one can't lose!) I can't say this is the same "candidacy and election" like I experienced in the communist system, but it makes me think, and it comes extremely close in my mind (except two parties instead of one)!

One more thought to the "**Freedom**" issue. I made a bold statement that "No one is really free!" in the capitalist society. I argued that the current capitalist society is the Level 2 emotional "freedom" and not the Level 3 Logical "freedom" based system.

The original founding of the USA happened because the founding founders were (using a nice word) "annoyed' by the "English King". So they declared in the "new document" that we all are created equal and should be free under God.

It is easy to understand why and where "freedom" came into their life. In reality, we have our boundaries as I described before and it is emotionally very disturbing.

Part 2 - Chapter 17
Philosophy

Philosophy, Section One (cont.)

So we created a word, "freedom", to lift our emotional status. Let us look at it closer — how is it working? At the time of the declaration of this new nation about 230 years ago, relatively few people were living in this land compared to today. They also had very little technical help compared to today's environment. It is very important, because if one sits on a horse and rides 1 or 2 weeks to get to the next populated small town, one can encounter an "absolute freedom" on the way. For that time the emotional experience reflected reality, because the traveler found very few disturbances out there, (other than some unhappy "Natives" for one's intrusion), and the natural surroundings with the amazing richness of the plant and animal world. So the "emotional feeling of freedom" seemed (and came close to), the "real freedom".

Time went by and a little more than 230 years later we bump into each other in every possible way in life. (Overpopulation?!) How "free" is one when one stands in line at the gas station, bank, grocery store, supermarkets, malls, ticket offices, etc., even in the amusement park waiting for a fun ride? In reality, freedom is gone! Our "relative freedom" is getting smaller and smaller every day. So what is that "Freedom" we are talking about every day? It is interesting that every day, practical people know the limitation of our "real freedom".

So, why do we feel free? The answer is simple. In this consumption based capitalist society, the offer of products and services seems unlimited. The continuous invention of more and more products and services, consumed by our emotional set up, makes us feel free. (Before I go further I have to confess that I am not an exception either!)

We all have our interest areas and there are countless emotional choices as "freedom" for us. Unfortunately for us (in the USA) the trend is that more and more nations want to take a bigger and bigger piece of the distribution of the products of our planet. (Growing trade deficit with many nations every month!) This will lead our standard of living (USA) to decline and the discovery of the people that we need a **new and better structure** for our society. I will introduce that in Section Four with the "Unequal", "Fair Share Distribution System".

Part 2 - Chapter 17
Philosophy

Philosophy, Section Two

<u>**Section Two**</u>, the analysis of the history of Philosophy. I will start my investigation with "**Primal Religion**", from the time of about 100,000 years ago when we (humanity, actually the science of anthropology) are carbon dating the first "Homo sapiens". (Actually the "primal religion" as we know dates back as far as 3 to 5 million years). We need to notice here that no other Philosophy of the history of humanity goes back 3 to 5 million years to try to investigate. Mostly because in the time of their writing they didn't have 20th century knowledge like we have today. (No words in any of the religious publications, like the Bible or Koran, nor about "primal religion" either — that proves my point again, that they aren't the "Science of the Sciences, as I defined Philosophy, they are only "thought processes".)

We have very little evidence about the possible "thought process" of the "primal religion" from that time, because no written documents, or recorded pictures, or recorded movies are available from that time of human history (Remember, as a realistic philosophy, I am looking for tangible evidence, that we can observe or recreate in our "lab"). The only evidence that we have to go on is some of the findings of their remains, tools, living areas, and some cave drawings. We know from this evidence that they had the Level 2 Emotional and the beginning of the Level 3 Logical / Reasoning mindset.

Some of those cave drawings represent (in our (humanity) best guess) an emotional / logical preparation for their hunting, or the recording of their successful hunting. Definitely we can see emotions in the pictures and the logical design of spears, bows, and arrows that they used for their hunt. The reality of the "mirroring mind" and the realization of the functioning spear or bow, that can "bridge" the distances between them and the targeted animal.

We can't call it identical, however, we can observe and investigate in our modern world the Australian aborigine's and North American Native's belief systems. They thought that all things are part of a "one great spirit world" and they are only appearing in different formats. They were very good observers (they needed it for their survival!) so they observed nature's differences.

Part 2 - Chapter 17
Philosophy

Philosophy, Section Two (cont.)

However, they viewed distinctions as bridges instead of barriers. They looked at all beings (plants, animals, humans), all things (water, rocks), and all elements (wind, rain) as their brothers and sisters.

We can see from that time that the "thought process" was based on their observation and imagination to explain the unexplained. Harmony with the environment and the respect of the environment was the main focus of their thought process. We can observe here that the "early humans" had no concept of the "One God", or "God" as we introduced it in the "Books" like the Bible or the Koran — though they did see a "higher power" in many things that they couldn't explain, like lightening, earthquakes, and tornadoes.

For example, natives believed that "if one sees a "dead man" walking in a tornado, that one dies!". It turned out by our modern weather observations (by filming it), that a formation, like a "walking man" (they called it "dead man") appeared in the "level 5 tornado". A level 5 is more than 240-300 miles of wind speed and easily can kill a human being, so their observation was correct as a final result, but the formation of the rotating clouds and dust by the wind have nothing to do with the spirit world or a "dead man" concept — it just forms a "walking man comparable figure" by chance, because of the forces of the wind that created the physics of the collisions of the hot and cold air masses.

One can see that the "**belief system**" started to develop at the beginning of humankind. They encountered many "Unknown **Bigger** Powers" that, in their mind, were translated to many "Unknown **Higher** Powers".

It is easy to see how our human Level 2 emotional mind helped us survive through the emotion we call "Fear". The human fear helped us stay alert and responsive to outside dangers (like lightening, earthquakes, tornadoes, hurricanes) to preserve our life.

It is easy to see also that this "Fear" could and can create a belief system, believing that if one pleases those "Higher Powers" (with one's behavior, like offering them something like prayers) who are controlling those forces, that one gets looked on favorably by those "Higher Powers".

Part 2 - Chapter 17
Philosophy

Philosophy, Section Two (cont.)

Our ancestors had, not having the knowledge as our today's science, to explain all those powers as the result of Energy transformations in the Matter, the only "logical?" conclusions that the "Higher Powers" must be governing all those events. I put a big question mark after "logical?" because their logical conclusion was actually an emotion based conclusion without having much logic at all. They didn't have a Level 3 Logical / Reasoning mind developed as ours is today. "Higher Powers" started to exist in the life of humanity and it still exists today, however, the ever changing science keeps influencing the belief systems and they keep changing as science points out, more and more, some of their weaknesses in their concepts and reasoning.

It looks like in the early development of humanity, the Level 2 emotional mind started to develop faster, and there was a slow beginning of the Level 3 Logical / Reasoning mind to follow. The next 80 to 90,000 years, we didn't see too many changes in the status of humanity. The next interesting discoveries came from around 12 to 10,000 B.C., in the land of Egypt.

We (humanity) discovered from their writings on the walls of the Pyramids, that they were fascinated with the afterlife (And this creates the ancestry for all our currently existing religions, "Higher Power" + "After Life Existence Forever"). At that time we encountered a more structured population in the current location of Egypt, India, and China, however, in all these nations today, a more current form of belief system exists, which began around 6 to 500 B.C. Interestingly, all the most widespread religions on Earth started around that time. Before I go on and create a timetable for the events around the 6 to 500 B.C. religions, I have to investigate an interesting "thought process".

Here, all the information is coming from observations (facts), but the "thought process" will be very theoretical in nature. Scientists discovered that the layout of some of the pyramids exactly mirror the Orion star formation of about 12 to 10,000 years ago. Interestingly, the early Egyptians were very advanced astronomers. They also had some beliefs that their passing away relatives somehow were going to the afterlife somewhere in the stars.

Part 2 - Chapter 17
Philosophy

Philosophy, Section Two (cont.)

Some drawings on their walls of the pyramids look like a character that is very similar to our "astronauts", as far as the "special helmet" over the head of the character. Scientists measured that if one divides the height of a pyramid into the sum of all four lines at the base of the pyramid, then we get the mathematical constant of "Pi". The next observation, that when humanity did an unmanned Mars mission, we were calculating the closest distance between Earth and Mars as they were rotating around our Sun. The "Solar System" has it's rotation around the "Center of the Milky Way Galaxy". These observations are facts that anybody can check in the records of our current sciences.

Too many coincidences with our current 20th and 21st century knowledge is coming from 12 to 10,000 years ago. It created an interesting theoretical thought in my reasoning mind — it almost looks like someone tried to send a message into the future. One can ask a question: How does this information come into the history of religions? One needs to remember that McKaneism, as a Level 3 Logical Philosophy, tries to analyze our existing facts to conclude or, in this case here, build a possible theoretical explanation for the origin of religions.

It is a very logical observation that the "Higher Power" can do "things" that are humanly impossible at that given time of human history. For example, if I would have appeared in 10,000 B.C. Egypt with a Helicopter, hovering over one of their pyramid building sites, all the people of that time would have experienced a fact based event (observed by their eyes) that they couldn't do. In their knowledge world, anything that "drops" falls down to Earth. How could something "Hover" in air without falling down? The answer: "It has to be a "Higher Power". Just like I argued before, in this example, they experienced a "Bigger Power" than themselves (in this example, a technologically more advanced power) but not a "Higher Power".

All this reasoning creates a very interesting theoretical question. Is it possible that 12 to 10,000 B.C. years ago that our Solar system was close to another solar system, that may have had an intelligent existence at the time who was able to cross space and visit our planet? (Like Earth and Mars in our unmanned visit?)

Part 2 - Chapter 17
Philosophy

Philosophy, Section Two (cont.)

They sure would have appeared here as "Higher Powers" maybe "hinting at the thought" of the "structure of the pyramids" ("Pi?") or the formation of "Plato" ("as the Orion star system?") or wearing a "helmet" as our current astronauts do ("picture on the pyramid wall?"). Could it be a "message in a bottle?".

One more interesting "legend" that was passed down the millenniums by the word of one generation to the other in Egypt, that somewhere in those mysterious pyramids and buildings remains a "Knowledge" hidden that would change the history, the thinking, and the future of humanity. I won't conclude anything here, I only introduced the group of existing facts connecting them with some hypothetical questions.

A sad footnote for all this before I go on, that all this belongs to "Humanity!" as I argued before, just like natural resources such as oil, or lumber on the planet, and in this case a few individuals' "bad character", like always, are sitting on the mysteries and advising their own government not to allow the exploration of those "possible" findings, because these 3 to 4000 year old sites, with their mysterious history, is a better income source for their own country. (Tourism.) Back to the same "stupid humanity" a few "gold coins" (Loser Capitalism) created instead of a "possible future altering (for the better) findings" for Humanity! Nobody can be certain what is or isn't hidden there, but it would sure be worth every moment of exploration to find out.

We can also see from this theoretical argument that the existence of a "Higher Power" beliefs can have many possible origins or reasons. That leads us back to our main subject of the religions. Seems to me that mankind was looking out for an answer in the "Higher Power", where mankind itself didn't have the answer. I will continue on into history, to arrive around 6 to 500 B.C where the surge of "thought processes" appeared in our planet.

One can find the facts from our scientific research that the "Thinking Process" of humanity increased substantially around the 6 to 500 B.C. time.

Many outstanding individuals started to investigate the basic questions of philosophy, like "Who and What are we?, etc.".

Part 2 - Chapter 17
Philosophy

Philosophy, Section Two (cont.)

Based on the answer that they gave, they formed a basic thought process, that later many others followed, believed, and extended on.

I can't answer the question Why, in that time frame (6 to 500 B.C.), did humanity start to get real excited about answering the major questions of philosophy? The only theoretical answers I have, are that maybe that was the time when the "quantity to quality" changes took place in the human mind. (remember "Directional Theory", simple to complex slow process!) The other possibility, that in that time interval in human history the population on our planet was growing and (as we know from some of these "founders") the societies were very unruly and often warring with each other. (For example, we know from the Koran that Mohammed argued in his "thought system" that his rules will bring peace and organization into the "social structure", that was very chaotic in his time.)

In summary, all these "philosophies" or by McKaneism "only thought processes" asked the same questions about ourselves and our possible "social structures". Interestingly, that time they all focused on only the Second Level (Chart 16) of the "social structures".

In the next page, I will introduce a timeline of our "Philosophers" as we (humanity) know from our scientific research, then I will return to extend my Chart 20 to make it more complete.

I won't analyze in every detail every "philosopher" but I will only introduce their main contributions (positive or sometimes negative) to humanity.

It would be impossible to fit this analysis in this section with that detailed a view and it wouldn't change the main point of the analysis, as we will see, that the process by the "Directional Theory" of humanity started to understand itself better and better. (Of course, with the dead ends of some "philosophies" as part of the "growing up process" of humanity).

The Chart 40 (extended Chart 20) will place all of these into the chart, who appeared in our timeline. The timeline starts at 100,000 years ago, so it won't be accurately proportioned in the page for practical fitting (as many possible "thought processes") on one page.

Part 2 - Chapter 17
Philosophy

Philosophy, Section Two (cont.)

Many times in my book I remind my readers that from the viewpoint of McKaneism, the "events", the "players", and "their thought processes" are very important, however, the timing precision of a few years ahead or behind doesn't make any difference in our human history or my analysis or my investigation. It is easy to see that I argued before, that one or another "thought process" appeared in history that we still can observe and experience today (like Christianity, Islam, Judaism, etc.) and it is highly irrelevant that some thought process appeared 10, 20, or 50 years earlier or later. Our scientists, as they know better and better with their discoveries, will put all these events in an exact timeline. Today, only those who argue on the timing of these "thought processes" (to contradict them) has no valid argument against their "thought processes" (not like McKaneism), they only pick on their timing to discredit their validity.

Here is one possible, scientific timeline of the previously existed human "thought processes":

Chart 40

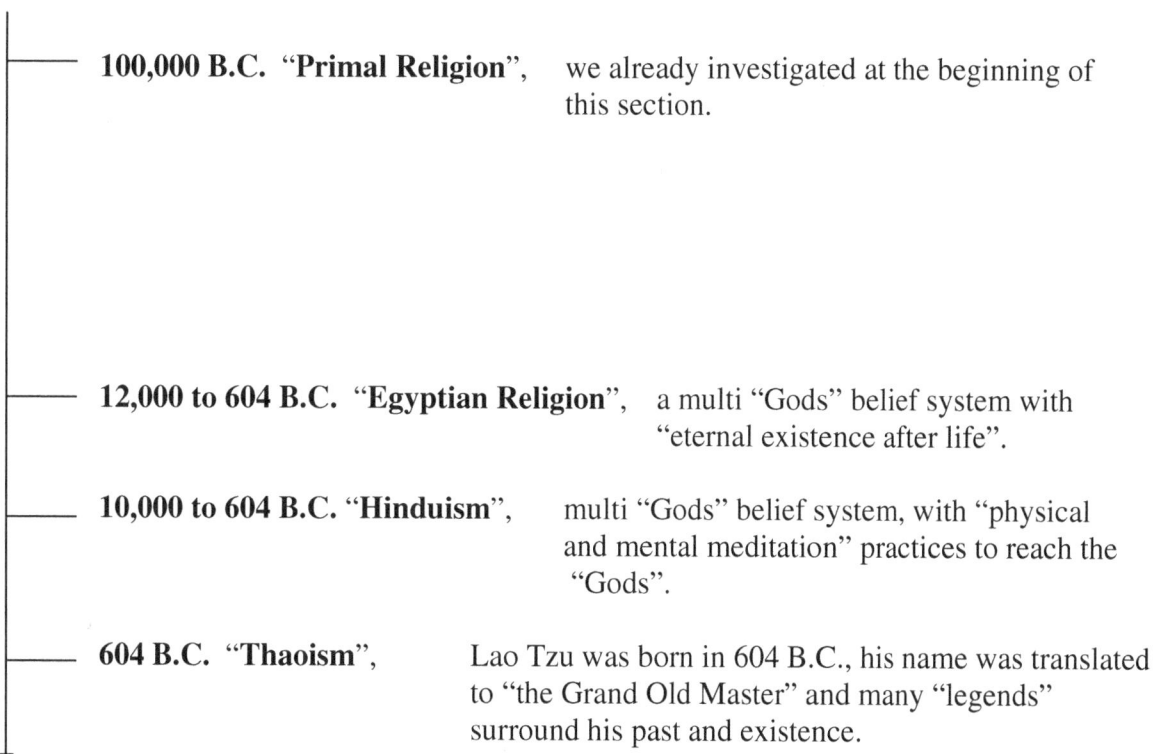

100,000 B.C. "Primal Religion",   we already investigated at the beginning of this section.

12,000 to 604 B.C. "Egyptian Religion",   a multi "Gods" belief system with "eternal existence after life".

10,000 to 604 B.C. "Hinduism",   multi "Gods" belief system, with "physical and mental meditation" practices to reach the "Gods".

604 B.C. "Thaoism",   Lao Tzu was born in 604 B.C., his name was translated to "the Grand Old Master" and many "legends" surround his past and existence.

Philosophy, Section Two (cont.)

Chart 40 (cont.)

585 B.C. "Milesians",   or Ionians in Greek (Thales, Anaximander, Anaximenes),

570 B.C. "Islam",   Muhammad, the prophet of "God", was born in 570 B.C. founding the religion of Islam (primarily means "peace")

563 B.C. "Buddhism",   Siddhartha Gautama who was born 563 B.C. originated Buddhism that from his answer: "I am awake" for the question: "What are you?" translated to the "Enlightened One" or the "Awakened One".

551 B.C. "Confucianism",   Confucius was born 551 B.C., the Chinese regarded him as "First Teacher", that is first in rank not first in time.

525 B.C. "Pythagoreans",   Greek (Pythagoras)

504 B.C. "Heraclitus",   Greek, idea of change (first in the relativity thinking) "all things in the *"flux"*.

490 B.C. "Atomists",   Greek, Leucippus (490 B.C.) founder,   **"Democritus"**   (460 B.C.) detailed elaboration.

470 B.C. "Sophists",   Greek, Athens (Protagoras, Gorgias, Thrasymachus)   **470 B.C. "Socrates",**   Greek, Athens (critics of Sophists.)

428 B.C. "Plato",   Greek, Athens, The first Philosopher with one of the most extended "thought process". He tried to cover all the elements of human existence.

384 B.C. "Aristotle",   Greek, attended Plato's Academy in Athens. Later he opened his own school, the "Lyceum".

342 - 204 B.C. "Epicureans", "Stoics", "Skeptics", "Neoplatonists".

4 B.C. "Christianity",   starting with the birth of "Jesus". (Official around 314)

0 ---

Part 2 - Chapter 17
Philosophy

Philosophy, Section Two (cont.)

<u>**Chart 40 (cont.)**</u>

70 "**Judaism**",   (70 C.E.) After the destruction of the Temple by the Romans, the Jews had rebuilt it, and started to focus on the study of the Torah. (Some sources originate Judaism by the birth of Abram around 1800 B.C.E.)

354 "**St Augustine's Christian Philosophy**",   imperfect reality, peace in eternal truth.

480 - 1200 "**Philosophies of the Dark Ages**",
480   **Boethius**,   high position, honor, suspected treason, executed (after prison sentence) in 524.
500   **Pseudo-Dionysius**,   tried to relate Christian thought to Neoplatonic.
810   **John Scotus Erigena**, argued: only one true reality, this reality is God.
1000-
1200   Preoccupation with proving God's existence.

1225 "**Scholastic System of St. Thomas Aquinas**", a theologian, relied on Aristotle, tried to provide logical proofs of God's existence.

1466 - 1724 "**Philosophy reflects the conflict between the force of Christianity and the growing force of Science**".

1466 "**Erasmus and Luther**", Erasmus critic of the Church, Luther a reformer
1469 "**Machiavelli**", analysis of the political behavior
1478 "<u>**Sir Thomas More**</u>", Utopia
1533 "**Michel De Montaigne**", classical Skepticism
1561 "**Francis Bacon**", tried to reform the philosophy and science in his time
1588 "**Thomas Hobbes**", a single method knowing matter, man, and the state
1596 "<u>**Rene Descartes**</u>", called the "father of modern philosophy"
1632 "**Baruch Spinoza**", exact knowledge of reality by his method of geometry
1632 "**John Locke**", founder of empiricism in Britain
1646 "**Gottfried Wilhelm von Leibniz**"   tried to achieve a reconciliation between Protestantism and Catholicism
1685 "**George Berkley**", set out to deny the existence of matter
1711 "**David Hume**", clearest and most rigorous formulation of empiricism
1712 "**Jean-Jacques Rousseau**", "The Social Contract", citizen, a civil society
1724 "<u>**Immanuel Kant**</u>",   Metaphysics, attempt to include all of reality in mechanical model

Part 2 - Chapter 17
Philosophy

Philosophy, Section Two (cont.)

<u>Chart 40 (cont.)</u>

1770 "<u>Hegel</u>", Absolute Idealism, Logic and Dialectic Process

1798 "Auguste Comte", Positivism

1806 "John Stuart Mill", Utilitarianism, ("Jeremy Bentham", 1748)

1818 "<u>Karl Marx</u>", Dialectical and Historical Materialism

1820 "<u>Fredrick Engels</u>", Manifesto of the Communist Party

1839 "<u>Charles Sanders Peirce</u>",   Pragmatism, three philosophers from the
1842 "<u>William James</u>"   USA, concepts, "Cash Value" of things,
1859 "<u>John Dewey</u>"   answer for "Does it work?" to define value

1844 "Friedrich Wilhelm Nietzsche", Hitler inspired by "Will to Power"

1859 "Henri Bergson",   Two of the Twentieth Century
1861 "Alfred North Whitehead", Metaphysicians

1870 "<u>Vladimir I. U. Lenin</u>". Leninism

1893 "Mao Tse-tung", Maoism

1872 - 1911 "Analytic Philosophy", main arguments, clarify the meaning of
                          language, don't build complete systems, "isn't
                          a doctrine but the activity"
1872 "Bertrand Russell", Logical Atomism
1889 "Ludwig Wittgenstein", "Philosophical Investigations"
1891 "Rudolph Carnap", Logical Positivism
1900 "Gilbert Ryle", "Concept of Mind"
1911 "John Austin", "A Plea for Excuses"

1813 - 1905 "Existentialism", most important element *existence*, argument,
         existence reserved for the individual human being, who strives, who
         considers alternatives, who chose, who decided, who committed himself.

1813 "Soren Kierkegaard", argued that Hegel lost the *existence* in his system
1859 "Edmund Husserl", connection between existentialism-phenomenology
1883 "Karl Jaspers", Being, studied from the point of *existence* philosophy
1889 "Gabriel Marcel", Being and the question "What am I?"

Part 2 - Chapter 17
Philosophy

Philosophy, Section Two (cont.)

<u>Chart 40 (cont.)</u>

1889 "**Heidegger**", focus on individual's thinking as an existing human being
1905 "**<u>Jean-Paul Sartre</u>**", Existentialism, one is simply that one makes of self
1905 "**Ayn Rand**", Objectivism, Introduction to objectivist epistemology

1953 "**Frank P. McKane?**", McKaneism (2003), The Foundation of Humanity, Philosophy the Science of the Sciences

I need to provide a short analysis to my Chart 40. First of all, I used the "**date of birth**" of all philosophers (I tried to be accurate based on my research materials), because it would be impossible to argue (or I can argue with everybody as long as we live) about the timing of their most important impact of their work on their societies.

Even from the historical view it is extremely difficult to place a point of impact on their work. One needs to understand that the nature of philosophy, that has each new thought process (for example like McKaneism) will be received with a lot of cautious investigation at the beginning, and then may or may not impact the thinking of the people as time goes by. We all know that we all need to grow up with our schooling before we can function in our societies, so all of these philosophers did the same. The impact of their thoughts came after they finished their school years, but their timing is just as many and different in variety as their life was in the time of history. The most important aspect of my Chart 40 (at least that I liked to communicate) is that they were existing, thinking minds in the history of humanity with a great impact on the "directional" thought processes of humanity.

Adding my name to my Chart 40, is only a pointer in time, that I think my Philosophy belongs in the history of thinking. I don't try to make a claim, however, I only hope that my "thought processes" will be worthy for the approval of humanity and humanity will put my name on that chart. One also needs to understand that there is great difficulty in building a chart like Chart 40. Most of the religions are alive and well today, and all of them were enhanced by the time of history.

Philosophy, Section Two (cont.)

As many ways as science tried to prove their invalidity, their practitioners picked up the old teachings and altered them and enhanced them to "make them fitting" for the today's societies. It is very hard to find the beginning or the originators of those that go earlier than 6 to 500 B.C. in the time table. I am also sure that I missed some of the influential philosophers in their times, however, not by choice but by the lack of visibility of them in the history of philosophy.

The other difficulty exists because philosophy always has been a **neglected science; more than that, it's not even considered a science** (that I am hoping will change with this book), to which is also true for the philosophers themselves.

I underlined some of the names of the philosophers in Chart 40, that means that those were the "thought processes" that influenced me most to compile, analyze, study, and build McKaneism as a philosophy. The list can't be complete without revisiting one more time the names that helped me create my thought process, and provided never ending interest in the ways of thinking. One needs to understand that interest in philosophy means that "every thought process" from the philosophers like "Aristotle" to the "corner gas station clerk" influence one's thinking. The person with philosophical interest never knows whose "small sentence or a few words" will "turn the light on" to connect the thoughts, and that can lead to the great discoveries of new "thought processes". One thing I can say for certain is that I had **"core influences"** by those names I will reiterate, and of course, the on going influence of every event I encounter and every person I talk to.

The **"core influences"** were coming from **Democritus** (the first one to think that everything was built from a small building block the "Atom"), **Socrates** (the philosopher that always found the way in every argument, to look at every possible angle of every possible reason), **Plato** (the first thought experiment with the "Allegory of Caves" pointing out the possibility of "brain washing people"), **Aristotle** (the one and only philosopher before me, who built a **"complete system of thought process"** a **Philosophy**), **Christianity,** the importance of the moral value system in society, (like the Ten Commandments).

Part 2 - Chapter 17
Philosophy

Philosophy, Section Two (cont.)

But, I don't think it has to be originated from God!, **Sir Thomas More** (in his book of the "Utopia", he describes a human community existence secluded from the natural environment, and doesn't interfere with a natural environment), **Descartes** (with a concept of thinking "cogito ergo sum", "I am thinking so I am", that lead to the self realization of the mind), **Kant** (with his argument that the laws of nature can govern the universe without any existing "Higher Power"), **Hegel** (one of the most complex thinkers in the history of humanity, reading his work extends the dimensions of one's thinking), **Marx, Engels, Lenin** (the so called communist philosophers, influenced me with their logical thinking, and the bold idea that the "structure of the society" can be designed artificially by humans).

I am positive that Marx was the "most complex thinker" in the history of humanity so far! His mind was "run by" the thinking way of the "Dialectical Logic", that one can read, enjoy, and understand from his writings (however, in the history of humanity, no one ever compiled the book of "Dialectical Logic" yet, as we have a "Formal Logic" for our mathematical language). And last, but not least, **Jean-Paul Sartre** (the importance of the individual existence, that help me formulate my own views of the "Attributes of Human Beings Theory).

Of course, a list can go on for many pages, but just a few more important influences from other than the area of philosophy: **Euclid** with his system of geometry, **Einstein** who extended the "thought experiments" to the possible limit in our time, and in my view the "second most complex thinker" in the history of humanity so far! (For example: (this is from Einstein as poorly interpreted as I can understand) a "thought experiment" that two trains are travelling parallel to each other and are moving with the speed of light.

Someone flashes a flashlight from one train to the other. What is happening with that light beam and how does the other person observe it on the other train? If you think it is simple, just think about it, that the speed of light is the fastest that one can go. How then does that light "catch up" with the other train that is also travelling with the speed of light?, etc.) One can see how complicated a "thought experiment" can get in Einstein's mind, and it gets even more complex as he puts them in the language of the mathematics.

Part 2 - Chapter 17
Philosophy

Philosophy, Section Two (cont.)

The philosophy of **Zen** from Japan, that seemed like a "materialist religion!?". The founder of Aikido, the "Great Master" **Morihei Ueshiba**, whose teaching introduced me to a possible way that one can discipline one's mind, and can create the harmony of "Body, Spirit, and Mind". Of course there are many, many others as I said in the beginning of my book. Every thought is extremely important for a philosopher, because one can never know which thought will connect to which one to form an enlightening moment in the struggle on the road to discovery of the "relative truth"!

Now, I will take some time to analyze some of the previous philosophies, and of course point out some of their major weaknesses in their explanations and reasoning. I will follow here Marx's recommendation that each "thought process" has to be viewed and analyzed "in their own historical environment". It is easy to be critical with any of them with today's knowledge, but it would be unfair to use it against them. I will only analyze some of their "illogical" logic, to point out the invalidity of some of their reasoning. Here, I am not picking on anyone, but I have to choose only a few out of the many, because the volume of this book wouldn't be sufficient to argue with all of them.

One needs to understand that in my "thought processes" of McKaneism, I learned a great deal from them, as the way of thinking and as their extensive knowledge base, however, my interpretation of their thought processes can point out that most of them were occupied to criticize something in their time of history, and with the one exception of "Aristotle", none of them built a comprehensive system of thoughts or a complete philosophy of their time. I can start with "**Aristotle**", who investigated every area in his philosophy from the sciences to politics in his time, however, he wasn't able to tie his great volume of information into a system as McKaneism does in Chart 26. I have to express again that he was the "most complete" philosopher in history, but even he didn't place "Philosophy as the Science of the Sciences".

One can see that most of the religions are built on the "belief system", that we are created and guided by "Higher Powers" or a "Higher Power".

Part 2 - Chapter 17
Philosophy

Philosophy, Section Two (cont.)

Somehow the religions shifted from the early "multi gods" concept (Primal, Egyptian, Greek, Roman Religion, Hinduism) to the "single god" concept (Islam, Christianity, Judaism). Of course, with the first "Materialistic" philosophy of Aristotle, the non-believer was born establishing atheism (being an atheist as we call all the non-believers in God today).

From that point of time philosophy was continued in two lines. **Line one,** with those who tried to prove the existence of God (with building a belief system in the individuals or with logical reasoning), and **Line two,** with those who tried to prove the non-existence of God with their logical arguments. One needs to understand that "Belief" is an emotional state of mind, and no logical argument (or light) can filter into those minds (A "Screen door" is replaced with a "solid steal door" to keep the light and many other things out).

Those who tried logical arguments for the existence of God, their logical arguments were very weak.

<u>For the first example:</u> let us look at the main logical argument of Judaism for the existence of God:

"In the beginning God…." From beginning to end, the Jewish quest goes for their understanding of God.   "Whatever a people's philosophy, it must take account of the "**others**". There are two reasons for this. First, no one seriously claims to be **<u>self-created</u>**; (**The first problem in logical thinking is here!**) and as they are not, other people (likewise being human) did not bring themselves into being either. From this follows that **humankind was issued from** (**The second problem in logical thinking is here!**) something other than itself".

I don't need to go any further, I already found a logical problem. In their **true** statement,they claim that "**no one seriously claims to be <u>self-created</u>**"; and **it is true, that no one can claim that**. As we know that by our observations and human science, we don't create ourselves, however, logically when one says "**no one seriously claims to be <u>self-created</u>**" then one "**<u>quietly</u>**", without speaking about it (hidden in the logical statement), **<u>"Assumes" that we have to be created!</u>**

Philosophy, Section Two (cont.)

(when you are "not self-created", that means you were "created" the other way than self" — you were just "created" "logically"!) We know from our logical, biological science, that the cells of one special reproducing male and one special reproducing female meet by chance, and **"that is how we humans begin"**! The **creation,** that by the definition of many, to have **"something come out from nothing"** doesn't apply to this process at all! So, this whole "illogical and unproven" argument is based on that **creation** has to exist.

So, we prove **Creation** with **Creation**! **That statement has no logical, reasoning validity that Creation, as a process, has to exist in the beginning of humanity.** Of course, we find the second logical problem from the same **"Assumed Creation Exists"** argument, because if creation exists (that they **Assumed** without proof) then that is where **humankind was issued from.** One can see that if we handle Creation as an **"Assumed"** and **"not a factual or proven event"** — this whole argument from the beginning is nonsense.

For the second example: let us look at the main logical argument of the Scholastic St. Thomas Aquinas for the existence of (Christian) God:

"Aquinas argued that all knowledge must begin with our experience of sense objects. The chief characteristic of all sense objects is that their existence requires a cause. From here he argued that the existence of these objects requires a **finite series of causes (Here is the problem in his logical thinking!)** and ultimately a First Cause, or God".

I find a logical problem in his argument here. The same kind of problem that we saw before in the first example that Aquinas "**Assumes**" the "**finite series of causes**" has to exist. Here we have no proof for that in his logic, **it is only an assumption.** So his argument is unproven. On the other hand, we can "**Assume**" that we are dealing with an "**infinite** series of causes" just as well as "**finite**", then we **can't arrive to the First Cause, (God)** because the **infinite nature of the series of causes.** So, we saw just as much nonsense in Aquinas' logical argument as the previous one in the first example.

We are back to the previous statement of McKaneism, that all belief systems are emotion based, and can't be proven logically.

Part 2 - Chapter 17
Philosophy

Philosophy, Section Two (cont.)

Logical thinking belongs to the Level 3 Logical / Reasoning mind; Emotional beliefs belong to the Level 2 Emotional mind. In the same line our philosophers were divided between science and beliefs throughout our history, constantly trying to discredit each other. They all made the same mistake, they made a final decision on the argument of "Matter vs. Mind" that determined and closed the mindset of the philosopher and his or her followers.

They all were based on science (**Matter first**, like Pythagoras, Democritus, Aristotle, Kant, Russell, Carnap), or belief (**Mind first** (or creator God) like Plato, Boethius, Erigena, Erasmus, Luther, Leibniz). The others, most of the time, left their religion out of the picture (other than one can't lose one's "screen door" to view the world) and focused on some "section or part" of the philosophy, analyzing psychology, or sociology, or political behavior, etc. (like Machiavelli, Bacon, Hobbes, Descartes, Spinoza, Locke, Hume, Rousseau, Ryle, Austin, Heidegger, Sartre).

One also needs to understand, as I argued before, that the Level 3 Logical / Reasoning and Level 2 Emotional state of mind always go together. It can't be separated, however, in a Logical argument one can control the emotions, at least temporarily, so it won't interfere with the reasoning mind. Finally in this section I will extend on Chart 20, not with the intention to place all the philosophies or "thought processes" into another chart, but to define the meaning of the organization of the chart based on the historic definitions of the historically assigned categories. I have no need to change these categories, because they are well defined and organized and the "thought processes" are based on their belief system or their arguments are based on belief or logical reasoning.

I will provide few examples also for better understanding, however, I have no intention to place all philosophies in this chart. If one likes a little mind exercise, one can easily do it, based on Chart 40, to organize the "players" into Chart 41. Let me investigate Chart 41 now as a final step in this section.

These are the existing categories as of the year 2003. Please note here, that **Realistic can't be Subjective, because reality by definition is based on objectivity.**

Part 2 - Chapter 17
Philosophy

Philosophy, Section Two (cont.)

<u>Chart 41</u>

<u>Categories of Philosophies</u>

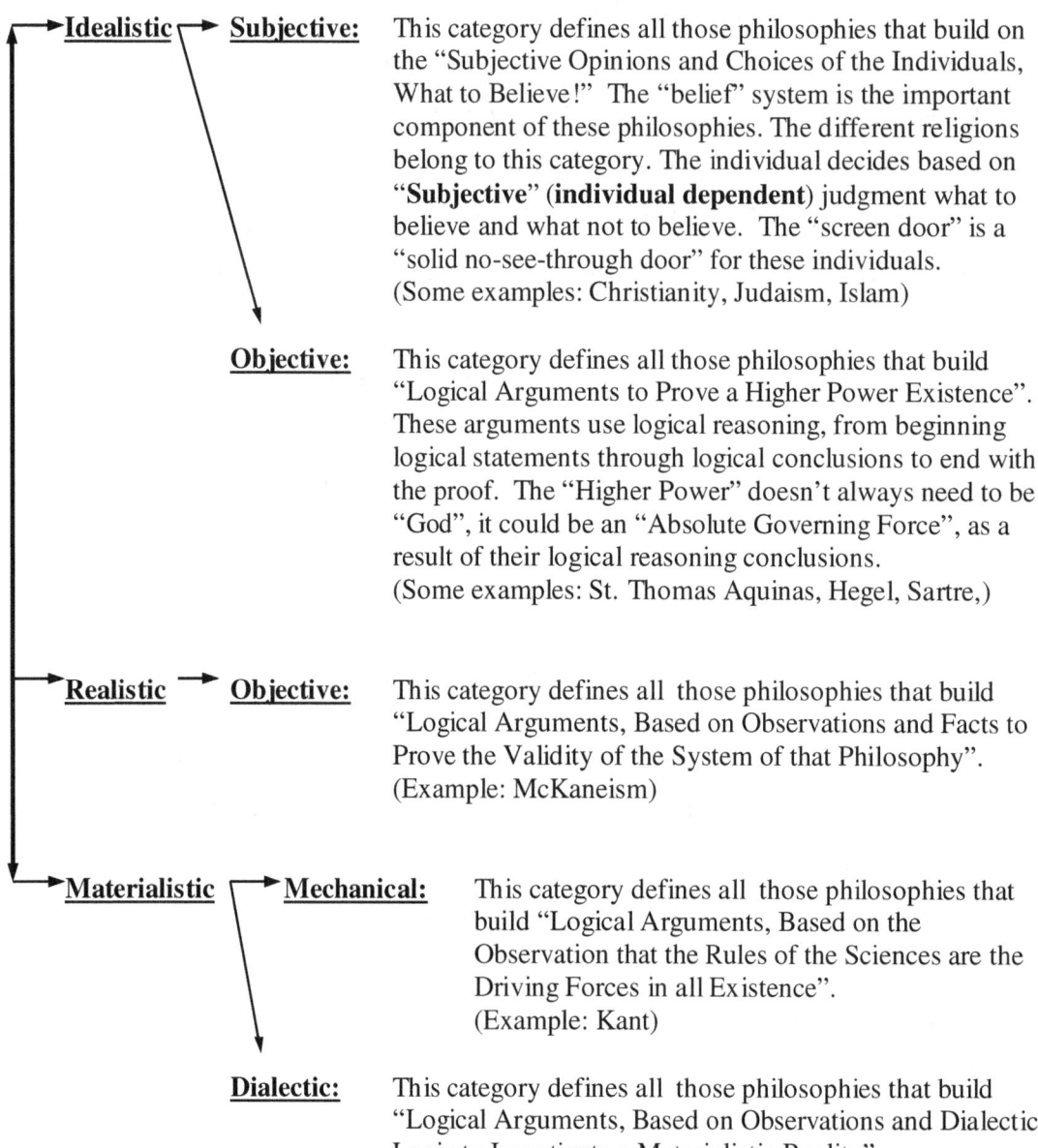

Idealistic → **Subjective:** This category defines all those philosophies that build on the "Subjective Opinions and Choices of the Individuals, What to Believe!" The "belief" system is the important component of these philosophies. The different religions belong to this category. The individual decides based on **"Subjective"** (**individual dependent**) judgment what to believe and what not to believe. The "screen door" is a "solid no-see-through door" for these individuals.
(Some examples: Christianity, Judaism, Islam)

Objective: This category defines all those philosophies that build "Logical Arguments to Prove a Higher Power Existence". These arguments use logical reasoning, from beginning logical statements through logical conclusions to end with the proof. The "Higher Power" doesn't always need to be "God", it could be an "Absolute Governing Force", as a result of their logical reasoning conclusions.
(Some examples: St. Thomas Aquinas, Hegel, Sartre,)

Realistic → **Objective:** This category defines all those philosophies that build "Logical Arguments, Based on Observations and Facts to Prove the Validity of the System of that Philosophy".
(Example: McKaneism)

Materialistic → **Mechanical:** This category defines all those philosophies that build "Logical Arguments, Based on the Observation that the Rules of the Sciences are the Driving Forces in all Existence".
(Example: Kant)

Dialectic: This category defines all those philosophies that build "Logical Arguments, Based on Observations and Dialectic Logic to Investigate a Materialistic Reality".
(Examples: Marx, Engels, Lenin)

Part 2 - Chapter 17
Philosophy

Philosophy, Section Three

<u>Section Three</u>, is where I will do the analysis of the **logical mind** of humans, (by McKaneism), that we (humanity) use (more or less) in our every day life to accomplish our logical tasks. (For example, designing a "blue print" of a house). One needs to understand that we (humans) weren't born with the developed logical mind. I represented in Chart 11, the three possible Levels of response (Level 1,2,3) of our human mind as an attribute of humanity. We also saw in the same chapter (Chapter 13) how our psychiatrists are able to alter our emotions and behaviors by chemically affecting the mind. We also know from our brain doctors that the mind has three layers and that corresponds to the three levels of response. In Level 1 and Level 2 all the responses are controlled by chemical compounds and electrical impulses in our brain. I am probably not too far from reality when I say that our Level 3 Logical /Reasoning brain functions have to be based on some chemical compounds and electrical impulses, however, we (humanity) know very little about how those connections function.

Most of our **Level 1 Reflex** functions of our brain are "invisible" in our human existence. Most of these Reflex reactions are "hard wired" into our brain, as basic attributes, and we have them from the time of our birth. (For example the control of our "heart beat").

The **Level 2 Emotional** functions create the "make us feel good or bad" chemicals, that controls a big part of our life. It is human nature that we all want to feel good most of the time and try to avoid feeling bad. Most of our important biological functions are controlled by our Level 2 Emotional response center of our brain. That is the middle part of our brain, mostly controlled by the chemical functions of our Hypothalamus.

For example, the "pleasure feeling" of good food does not just make us feel better, but the reinforcement that we want to feel the same pleasure again and again helps our survival by repeating our good food consumptions every day.

Now we arrived at the **Level 3 Logical / Reasoning** mind, that will be our focus for our investigation in this section. The highest level of brain function (in our gray area of the brain), but it isn't totally independent from our Level 2 Emotional influence.

Part 2 - Chapter 17
Philosophy

Philosophy, Section Three (cont.)

One interesting observation I had about the word called "Intuition" that comes into the picture for scientific discoveries. Myself (as a computer programmer), I worked with logical instruction sets and tried to translate the human language problems into the language of computer for faster or repetitive solutions. Occasionally I got "stuck" with the translation, and had to take a break from the problem. After a "feels good" dinner or exercise, with no warning, the solution just "jumped in" to the right place in my mind, so I was ready to finish my program and the problem is solved. I was wondering many times, and I still can't answer the question, of: "How does my "Feel Good" emotions help me solve a logical problem? I can't answer the question, but this observation of the reality in life, tells me that our Level 2 and 3 mind works together and can't be truly separated. I argued before that we (humanity) didn't discover the Logical mind totally, but we have through our observations and thinking some idea, that is discovered through history by philosophers and scientists. I will investigate and extend on some of their thought processes.

To talk about something "**logical**", that means in our language that from the "**Basic Starting Point A**" taking "**Logical Step B**" then "**Logical Step C**", "**Anyone!**" can get to the "**End Result D**"!

The four main important areas are here in this logical thought process example, that need to be logical. The "**Starting Point**" of reasoning, the possible "**Steps**", then it is for "**Anyone**" who travels on the road of logic, (no discrimination here) and the final importance of arriving at the "**Result**"! For example: If "Anyone" wants to visit me at my home, one needs to "**Park the Car**" (Starting Point A), "**Come Up 10 Steps**" (Logical Step B), "**Knock on my Door**" (Logical Step C), and I will open the door "**Greeting the Anyone**" (End Result D).

One can see that this example describes all the conditions of the logical thinking or logical thought process of this event. If I give these directions to anyone with a car, that person will find me following this logical thinking.

Part 2 - Chapter 17
Philosophy

Philosophy, Section Three (cont.)

(Of course one needs directions to my parking lot, but I don't want to start an example from "when you wake up in the morning", however, that would work just like this except that it would be three or more pages to read through).

The main concept here is that logical thinking is for "**Anyone**" who wants to use it. One needs to understand that we aren't born with a logical thinking mind. We need to go through many years of schooling and education from our parents, teachers, and all the other influences around us to develop the **Level 3 Logical / Reasoning** mind. One can remember from my Chart 29, Chapter 13, where I argued that the "**Triangle Q**" defines the complexity of comprehension and capability of logical thinking of the individual. In extension of that chart, one can see that the "**Triangle Q**" is only correct for "**Three Knowledge Arrows**", however, the real representation is a "**more complex object**" with as "**Many Arrows**" as the person's knowledge areas are. This information can lead to the measurement of one's logical thinking capability. This is where more education (higher education) comes into the picture.

One more interesting and very important and powerful point I have to make here, that the logical / reasoning thinking as we saw in this example is "**Emotion Independent**". We can have our emotions to "feel good" about the result or our capability to think logically, but the step by step logical process "**Has No Emotions In It**". For example: One can be happy or sad, but if that one doesn't take "**Logical Step C**" to "**Knock on my Door**", I won't open the door, so that one won't get to the "**End Result D**", that is the "**Greeting the Anyone**". (previous example)

Now, as we did see a small logical example from our every day life, and I defined a logical thought process through this example, we can move on to discover the different types of possible logical thinking, and how they are related to each other. We also will see the type of logical statement that we can define with our logical mind. In the investigation of "**Formal Logic**" I will bring in some of the Analytical Philosophers, and then Marx for the discoveries of the "**Dialectic Logic**" way of thinking, then I will extend this process with my own thinking using the "Directional Theory" to introduce a "**Multi Value Logic**" system.

Part 2 - Chapter 17
Philosophy

Philosophy, Section Three (cont.)

There are three types of logical thinking that I will investigate — they are thinking with "**Formal Logic**", "**Dialectic Logic**" and "**Multi Value Logic**". Any human mind is capable to do it, if trained for it. The next interest area, that I investigate, is the possible "**logical stream process**" that we saw in my introductory example going from point A, Step B, Step C, End Result D. I called it a "logical stream process", because the "**stream**" of steps needed to be followed to reach the destination. I will investigate some possible rules of this "stepping" process. Finally, in this section, I will analyze the "Rules" around us (humanity) from the point of view of the "**Specific**", "**Interval**", and "**General**" existing rules and their nature.

"**Formal Logic**" or "**The Logic of Mathematics**", that is used by many scientists and philosophers in their arguments to describe the universe that we are living in on a logical basis without the need for the existence of God. Kant (in his metaphysics) in his mechanical model, using the formal logic of mathematics, came to his conclusion that "Pursuing this method, science would have no need for, or could it account for, such a notion as freedom and God." Russell the "Logical Atomist" said "the kind of philosophy, which I call logical atomism, is one which has forced itself upon me in the course of thinking about the philosophy of mathematics."

McKaneism's argument about the "**Formal Logic**" is the following. The formal logic is a "**Two Value**" logic system, that in a given point of time can decide that a value is "**True**" or "**False**". Remembering my previous example, one can say that from the "Starting Point A", one takes "Step B" is true or false. If it is true, then one can take "Step C", that is the next step in the "logical stream process". If it is false, then "Step C' can't be taken, and our "Logical stream process" breaks there.

One needs to understand that without "**Come Up 10 Steps**" (Logical Step B), one can't "**Knock on my Door**" (Logical Step C). (I hear the smart one's argument that one can throw a rock to knock, however, that is not "**Knocking**"! This one has a huge problem with understanding the importance of the definition of words in a logical argument!)

Part 2 - Chapter 17
Philosophy

Philosophy, Section Three (cont.)

The next important feature of "Formal Logic" is that it's "true" or "false" valuation is viewed in the **"exact point of time"** existence (like a picture or snapshot). One can see here that a logic system in the snapshot mode may not describe our ever changing world (more like run as a movie). "Formal Logic" is one of the ways our logical mind is capable of thinking.

Here is one example of some problems with "Formal Logic": 4+4=8. We all know this. This is a mathematically "true" equation. One needs to realize that lots of information is in this simple equation. If one tries to use mathematical logic for proof in their investigation, one needs to understand that we have a problem when we introduce time. One can argue that 4+4=8, regardless of the timeline. However, here is a "thought experiment" to make one think. "I take a spaceship and visit another civilization. I present them with the language of mathematics that 4+4=8 and realize that they are clueless. How can this event happen? For example, this civilization is living with a "binary system". In their binary system, they have no digit of 4 or 8. This is my 10-based number system (numbers from 0 – 9). Their system has only 0 or 1. Here I am having trouble to communicate with my "math".

If they have no concept of my 10-based number system, I can't communicate with them. That leads us to my **first conclusion,** that one needs to use a logic that is understandable to others for establishing communication on the logical level. For another example, if I need to use calculus to prove one of my points, that logic probably won't be understood by 70 to 80 percent of the population on this planet (I can be off with my number here, it is only a guess, not a statistic, but I'll take anybody's statistic for my point!) To continue my thought experiment "If I write for my visited friends 0100 + 0100 = 1000, then they will understand my mathematical equation (Assuming they have a concept of "+" addition). That is the 4+4=8 in the language of "binary based numbers". Now we can communicate because I had the knowledge to translate my number system (10 based) to their number system (2 based). One can see how complicated logic can get even in this very small example.

Part 2 - Chapter 17
Philosophy

Philosophy, Section Three (cont.)

This example in math is like an ant compared to rocket science. Our mathematicians can come up with **"Way More Complicated"** formulas! That leads to my **second conclusion**, that one needs an **"overlapping knowledge base"** in the **"world of logic"**, to communicate through logic (Remember, Chart 30). Sometimes I'm watching people argue and I realize that one tries to make a "logical argument" and the other "fires back" with "flaming emotions". (Try an argument with anyone in the religious belief area, and you will realize that you are attacked from their "Sensitive Emotional Foundation", and this brings the "Fury" out of them. Try "Darwin vs. God" and you will see what I am talking about!)

For another argument, that formal logic can have trouble with time, I provide this argument. I can view a "parked car" and make a statement, based on my observation, that **"the car is parked at the number 7 spot"**. If someone is standing by my side at that moment, say at 8 P.M., this statement will be logical, simple, and **"True"**! (That is a "snapshot" in time!) However, if I say that to someone in my house at 9 P.M. and walk back to the parking area by 10 P. M. with that someone and the car drove away from the number 7 spot, I made a **"False"** statement, because **"no car is parked at the number 7 spot"**, now. So I was **"True"** at 8. P.M. and I was **"False"** at 10 P.M. So the validity of my logical statement depends on the timeline of the observation. (One can see that no emotion is involved to make a statement about a parked car! Simple, logical statement based on observation!)

That leads us to my **third conclusion**, that "Formal or Mathematical Logic" can have a problem with events observed and argued in the timeline.

If one thinks it is a very meaningless argument or investigation, that one needs to start listening to people's arguments to see how many times we fall into these traps of the "Formal Logic" in our logical reasoning.

The next level of Logical / Reasoning thinking we all can use is the **"Dialectic Logic"** in our thinking mind. We see the problem with the introduction of a timeline in the "Formal Logic" area. "Dialectic Logic" solves this problem to view all things in an ever changing, event driven environment.

Part 2 - Chapter 17
Philosophy

Philosophy, Section Three (cont.)

It is proper with "Dialectic Logical" thinking that my statement is "True" at 8 P.M. and "False" at 10 P.M. It doesn't make a liar out of me, just a simple acknowledgment of that change in reality with the change of time (I am referring to the previous "parked car" example).

"Dialectic Logic" introduces a few more elements in our logical thinking. In the events of the timeline we need to realize that we can have a "**beginning**" and an "**ending**" event, where the "**ending**" can be a new "**beginning**" in the timeline forming a "**stream of thoughts**". Marx argued that we need to view "events in their historical environment to reflect the proper logical way of thinking, reasoning, and concluding on them". For example: If I look at Democritus' philosophy, who said that "all things must be built from one small undividable element that he called Atom" and I say "this dude was stupid", I am missing Marx's point. In his time, about 2500 years ago, he was a genius to come to this conclusion with the available knowledge and tools of observation. One can see by this logically thinking example how Marx's observation and conclusion is correct, and how it is working in his "Dialectic Historical Timeline". We know today (our science discovered it) that the "Atom" is dividable.

For us today with our technology (cyclotrons, electro microscopes, etc.) it is much easier to investigate the "Atom", than it was possible for Democritus. One needs to realize that about 2500 years of the human growth process happened between the two analysis events of the atom in time (in the example in here).

That is the argument about the view of events in the timeline. One more interesting observation that I introduced here with using the words "beginning" and "ending", that many philosophers before me argued about, that it is the nature of things that we have a "**Cause**" and "**Result**" where the "result" can be the "cause" for the next "result". (Here we can find words used for "**Cause**" like "Cause, Reason, Motion" and for "**Result**" like "Effect, Result" used interchangeably accordingly).

Part 2 - Chapter 17
Philosophy

Philosophy, Section Three (cont.)

Marx used his "Dialectic Logic" in his head without anyone ever being able to write a "Dialectic Logic" book, as we have many of the "Formal Logic" or "Mathematical Logic" publications. One more interesting Dialectic Logical observation is coming from Marx, that the "**dual**" forces of events drive the process of "**progress**". He was referring here to an argument that the "class" warfare — the fight between the "Poor Class" and the "Rich Class" — is the "**dual**" force and this leads to the process of changing the structure of the society. One can like Marx as a Communist or one can dislike him for it, but with the introduction of the "Dialectic and Historical Materialism", as a philosophy, and the "Dialectic Logic" in his writings, he definitely was one of the most complex thinkers (I think "The Most Complex") in human history. One needs to remember that "Complexity" is not always related to the levels of volume of the knowledge base.

I can't prove it, but I think "**Formal Logic**" is a subset of the "**Dialectic Logic**", which is a subset of the McKaneist "**Multi Value Logic**". One can see that the Marxist "Quantity to Quality" change dialectic logical argument is a subset of the McKaneist "Directional Theory" — the "Simple to Complex Slow" directional process (Chart 23) that Marx applied for his social change theory in his philosophy. This observation leads us to the next possible way of logical thinking, and that is the thinking with the "**Multi Value Logic**" thought process. This thought process simply applies the facts that in my observations, in reality, we can have multiple "True" and "False" conditions in any event in the ever changing process of the given timeline. That thought process corresponds to the "Directional Theory".

The "Multi Value Logic" is a straight forward observation-based conclusion from the "Directional Theory". In the process of reality, that was described by the "Directional Theory", the reflecting, mirroring mind has a "Directional" or what I call "**Multi Value Logic**" — that means that **"one can have multiple "True" and/or "False" conditions in a directional event in the continuously changing directional process in the continuously changing timeline"**. The best example I think of is when a river runs out of it's bank.

Part 2 - Chapter 17
Philosophy

Philosophy, Section Three (cont.)

The river takes on a "simple to complex slow or fast" directional process that I can describe with the "Multi Value Logic" thinking. Based on the minute by minute changing conditions of the situation, I can be asked to answer "What is going on in certain flood area"? My answer can be "True" in one minute and "False" in the next because of the ever changing reality, that I try to mirror in my mind. The "beginning" and "ending" (cause, result) process gets replaced here (in the above example) with the **"beginning"**, **"progress"**, **"ending"** process (cause, progress, result).

One can see that if I use a "point of time event" in my "progress" to be a "beginning" (cause), then I can end up in a "Multi Value Logic" "stream of thought" process. I can have "True" and "False" "multi value logic" in the following (Chart 42) stream of thoughts:

Chart 42

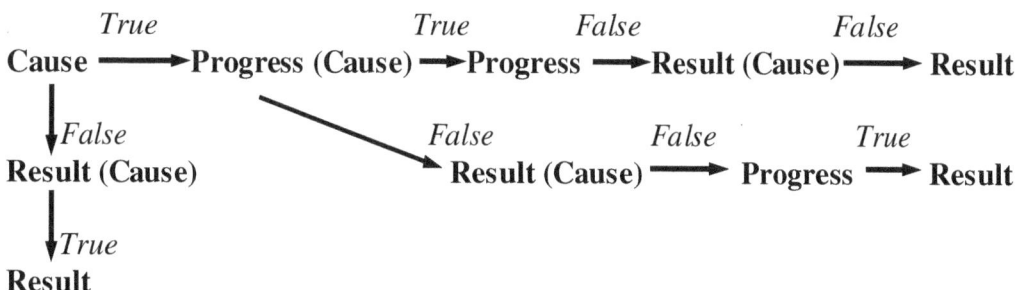

One can see that the variety of "true" and "false" values is unlimited in the logical thinking mind. One also can see that the result can be a cause of the next event, and the progress in any point of time can be a cause for another progress or result. This logical stream of thoughts defines our logical thinking that becomes observable through our every day communication in our language. It can be another barrier of communication between two individuals.

Part 2 - Chapter 17
Philosophy

Philosophy, Section Three (cont.)

Here we can see that we identified some problems in the area of human communication:

1. No overlapping knowledge area.

2. No overlapping understanding of words or their different definition.

3. No overlapping logical thinking.

I am referring here to each individual's "3D Object" in my Chart 29, that needs to have those three overlapping areas if they want to establish meaningful, reasonable debates and/or logical communication.

The next important aspect of our logical thinking is the definition of the "**Specific**", "**Interval**", and "**General**" logical rules.

This is the area where most of the misunderstanding can happen in our communication of our every day life. For example: I can say that: **one** man drives a car, so **all** men drive a car. Here, I went from a **"Specific" observation** (**one** man drives a car) to a **"General" conclusion** (**all** men drive a car) in my logical statement, but **my conclusion is incorrect!** Logically, from the "beginning" (cause) statement that "**one** man drives a car", I can't conclude the "ending" (result) statement that "**all** men drive a car". I only saw one, in the "beginning" (cause), so the second may drive a truck, and the third may drive a bus. That will prove logically the opposite that "**Not** all men drive a car".

One may think it is playing with words, but one needs to realize that serious logical arguments can't be solved if people don't understand the rules of logical thinking (it's not too relevant when one argues about some basketball scores, however, one better get the logic right when two nations have an argument and they are just about to start a nuclear war!) In the formal logic of mathematics the thought process is that, if a "Statement" is "True" for "**the first one**", and then the same "Statement" is also "True" "**from n to n+1**", then it is also "True" for "**all**"! One can see that it is fairly complicated to be correct on generalization.

Part 2 - Chapter 17
Philosophy

Philosophy, Section Three (cont.)

One can see that I can go from the "Specific" to the "General" or from the "General" to the "Specific" rules, and this investigation can be very complicated. Instead of doing that, I will have a few examples for these logical rules, without going into very lengthy logical explanations of all of them:

<u>For example</u>: The "**Specific**" rules always exist in a well defined specific time and space environment. If I use a half teaspoon of salt in my soup at cooking time, that is my "**Specific**" rule of my house. My friend may use a full teaspoon of salt in his soup, that is his "**Specific**" rule of his house. In reality, these "**Specific**" rules give us our own individual freedom within our own boundaries (like within our own house).

<u>For example</u>: The "**Interval**" logic rules usually exist in a given time interval. If I say "my dog is a good dog", that is an "**Interval**" rule statement. This statement is correct only for the "**Interval**" of time as long as I have my dog. If I give my dog to my friend, because he likes him very much, my statement has no valid meaning anymore because I don't have a dog. The "**Interval**" rule expired with the end of my **interval time** of ownership.

<u>For example</u>: The "**General**" logic rules exist all the time in the life of humanity. They are very powerful because they exist with or without our human approval or disapproval. If I say that the rules of gravity **always** apply **to all** individuals, I made a "**General**" rule statement. One can say what about weightlessness in space? One needs to understand that weightlessness is coming from the situation when the gravitational forces of Earth and Moon even out each other, or the orbiting spacecraft generates an "equal to" the gravitational force in the opposite direction to Earth. In our current existence we don't know any situation, when any event can happen without the influence of the rule of gravity. It is always out there and working, and that makes it a "General" rule.

The final logical rules in this section are the rules of the "**<u>Absolute</u>**" and the rules of the "**<u>Relative</u>**". One can see how complicated our reasoning can get with all these existing rules of logic working together in the same time in our logical mind.

Part 2 - Chapter 17
Philosophy

Philosophy, Section Three (cont.)

The rules of the "Absolute" represent an always existing permanent rule with no change in it's nature. In the human religions (Christianity, Judaism, Islam) the existence of God is an "**Absolute**" always and forever existing "Higher Power". The thought of "**Relative**" rules started with the Greek "Heraclitus", who was first watching a river and made a statement that "one can't step in the same river twice". In this statement he tried to explain that for the time period of the second step, the river will change, so it won't be the same river. He concluded that "all things in the *flux*". That was the first introduction of the relativity in things. If one remembers my Chart 22, it is easy to see why the constant change of things is based on the constantly changing energy level of things.

These two logical thinking rules created the most problems in the history of humanity. The criticism of the "**Absolute**" rules, that can lead to an Absolute existence, offended the basic foundations of most religions. The "Absolute" existence of the "Higher Power" goes further to introduce an "Absolute" Heaven and Hell — an "Absolute", always existing, never changing environment where the "well behaved ones" get their eternal reward, and the "bad behaved ones" will get their eternal punishment. These "logical? statements" totally contradict all our current observations and sciences. I put a question mark after logical, because even if it seems like a logical statement for the speaker, these statements are proven by the **emotional belief system**, rather than all those logical rules, that I described in this section.

They don't have any logical observations (For example, nobody ever had a picture from Heaven. It should be a picture with all the happy faces). So the first step is belief (that is emotional not logical) and the proof is going from there. A building built on quick sand.

The "**Relative**" rules are very important and alive in our human existence. We know it from our daily life, that every event is going in a time table, and that time table makes it relative. Any time one makes a statement we can see that it is compared to something, that makes it relative. The most important element here is the time. I can say Joe is taller than Jim, then I relate the two individuals heights to each other. I can't say that "Joe is the tallest" because what it is compared to is what is relative.

Part 2 - Chapter 17
Philosophy

Philosophy, Section Three (cont.)

That statement could only be true in "**Absolute**" terms, but I can say an absolute logical sentence about that. Introducing time the validity of the "**Absolute**" goes away, and the "**Relative**" nature of the statement takes place. If I say "tallest", I have to say relatively compared to what!

These are the rules that build a system of logical thinking that I call "Realistic Logicism", that one did read here in Section Three. It takes a long time experience of learning and practicing for one to use these rules in their every day life correctly. I did build my realistic philosophy McKaneism on the foundation of these logical rules, using them in my observations, reasoning, and conclusions.

Part 2 - Chapter 17
Philosophy

Philosophy, Section Four

<u>Section Four</u>, will analyze the most sensitive topic of my book, that is a "**Value System**". Before I continue, I must give a short introduction of the topic and the way I will analyze it. The first response that one can get for this topic is: "I am OK with **MY Value System**, thank you very much!" And then the next step is to be "**very offended**" if one doesn't drop the subject! **But, I can't!**

I will make a "bold statement" here again (as many times before), that we (the individuals of humanity) "don't have **Our Own** Value System", instead we all have a "**Learned and Constantly changing (Relative) Value System**", that reflects the values of the Society and the Philosophy that those Societies believe in. Let me investigate the "General", "Interval", and "Specific" logical rules of the "Value System". (I introduced those logical rules as part of the Realistic Logicism system in the previous section, and no surprise, they are coming around again!)

The "**General**" logical rule is that society can't live without a "Value System"! The peaceful interaction of individuals requires a set of accepted values to be able to function as a Society. (When "Value Systems" collide, then a non-peaceful existence takes place. WWII for example!) The foundation of the "Value System" in the Society is coming from Philosophy all the time. One needs to realize that a "**new born baby has No Value System**". A long growing, nurturing process has to take place to develop a "Value System" in one's life. The "**General Guiding Rules**" coming from Philosophy! (For example, as we will see later in more detail, in the USA we have a "Pragmatism, Existentialism, +1 Religion **Combination**" for our (USA) "Value System"!)

If one wants to argue about our Constitution, one is correct that our Constitution is a "**Great**" part of our "Value System" (USA), however, all those thoughts, rules, and beliefs were coming from the **Philosophy of our "Founding Father's"**! (So, philosophy is always involved in creating a "Value System"!)

The "**Interval**" logical rule of the "Value System" is that it is changing by time, and as a new Philosophy establishes a "better!?" (in quotes because it is very relative!) system, everybody buys into it and the society starts to live by a new "Value System".

Part 2 - Chapter 17
Philosophy

Philosophy, Section Four (cont.)

For example, around 1200 the Feudal Europe mostly lived by the "Value System" of Christianity, and then in the 20th century we added Pragmatism and Existentialism to it!

The "**Specific**" logical rules are the ones that make the "Individual's Offended" as I outlined in the beginning of section four. These are the values that each and every one of us can develop and practice as our own, and we all fiercely defend when one tries to "Step on our toe!". (For example, I am brushing my teeth five times a day, that is my "Specific Value" in my own "Value System"!)

The "Specifics" are very emotional values and these are the values that give one the "freedom feeling" in one's life. For another example: "I can go to the ball game twice a week in the season, because I am free and I like it!" That is another "Specific" rule, that one's individual based value is the liking of baseball! Try to tell this person that he is wrong, and a fight is in order, for one to defend one's values. (More precisely, one's "Specific Value System"!)

I am sure one can start to see the complexity of the "Value System".

Before I investigate our current (USA) "Value System", we need to see a few more angles of and in this topic. (One can see that each country can have their own "Value System" and they can oppose each other! One option is to "fight it out" with force, but the better solution (my recommendation) is to create an artificial "Best" working "Value System for Humanity" and let everybody "merge" into it, because it will be in their best interest! A peaceful merge to One Nation!?)

The "Value System" is artificial and "man made", other than the "Natural System". (remember Chart 17, #2 to #6! #7 won't be an exception either when it gets discovered by the people!) Anything "man made" has an "agenda!". The "goal" is to gain control over the other man! (For example: in the religion of Islam one has to go down on one's knees to pray, that is a behavioral control — or the "Ten Commandments" in Christianity that tell one how to live, what is right and wrong!) In that aspect #7 (McKaneist) is an exception! The objective of my "Value System" is to promote the "Survival of Humanity and the Fairness of the Society"! I am not interested in any personal "power control" over anybody.

Part 2 - Chapter 17
Philosophy

Philosophy, Section Four (cont.)

The "Value System" is based on the "social structure" of the Production/Distribution system. (I guess some of my advanced readers already figured this out on their own!) (For example: the "Slavery Structure" values slaves, the "Capitalist Structure" doesn't value slaves, more than that, they place a "Negative Value" on Slavery!)

One can't analyze a "Value System" without "Offending almost everybody" in the Society, one way or the other in the existing "Value System".

One needs to realize that for me to introduce a "**New**" and (I am sure) "Better Value System", I need to put the existing "**Old**" system under the microscope and analyze it. I will try to be as logical as possible, but it is very hard to leave out my emotions, because it is a "very compassionate topic" for me, especially the "Survival of Humanity and the Fairness of the Society" part!

Having said that, I want to say here only one time in the beginning of this section, that **I am most sincere, in the bottom of my heart as I write this section.** Sincere, because my interest here isn't a "power control" gain, but to secure and enhance the future of humanity. **I fully understand the difficulty and weight of these arguments. I am writing here with very great concerns of the non-existing "Self-Defense" of the "human race", that are jeopardizing our survival as humanity!** Any individual can understand individual self-defense but to extend to the "**Level of Self-Defense of Humanity**" is just as difficult to an individual as it is difficult to understand the speed of light without traveling it, because these dimensions are larger than our individual understanding of them.

I will argue to build a "Logical / Reasoning Framework" instead of today's existing "Emotional Value System". I will also place the human emotions into the system, as I argued many times before our "Emotions" are just a part of us as our "Logical Thinking".

It is easy to see that **I can't create a "plan to follow for humanity". It would be foolish of me to think** (like many others did!) **that I can design a "working, logical, artificial value system" for humanity by myself.**

Part 2 - Chapter 17
Philosophy

Philosophy, Section Four (cont.)

My objective is to introduce weaknesses of the current, "Old" existing "Value Systems", (mostly focusing on USA, because my hope is that my nation will lead the changes of the "Planet Earth") **and outline a large part of a possible "New" and designed "Value System".** Some of these arguments are "Theoretical", but not impossible, and definitely viable, doable alternatives for a "Value System" of the future.

Let me start with the definition:

"**Value System**" **is a summary of all written and unwritten rules that one nation as a Society lives by, governs itself by, and functions by in their every day life with all the unwritten habits and values of the Individuals included in it.**

One can see a "**Very Complicated and Complex**" factual aspect of this definition. (It's complexity is coming from "Reality", not by my personal "making". I didn't create the existing "Value System", I am just living by it an observing it!, and of course trying to enhance it!) The complexity is coming from the notion that it is a "**combination of written and unwritten rules**", "**constantly changing in time**", "**nation by nation based**", (should be "one humanity as a nation based", one of my arguments was for the future) and the "**unwritten habits and values of the individuals**" also involved in it. **These observations lead to a conclusion that a "Value System" is an ever changing existence with the interaction of the "Individual vs. Society" and different "Society vs. Society" on Earth.** The arguments here will "stir up" the emotions of many of my readers. Some will feel very positive and agreeable with my thoughts, and some will be extremely disturbed and angry about them. One needs to understand the following facts.

The "Value System" is built on the First Level (Production-Distribution System) and Second Level (Economy, Law, Ethics, Morale, Aesthetics, Religions, Art, Politics, Education) as I described in Chart 16. For this section I will stay in the USA, however, the interaction of the "**Individual vs. Society**" **and "Society vs. Society"** are constantly existing in any "**Value System**" because of our "**Global Village**" (Earth) existence in the 21st century, which is thanks to our advanced communication systems, like the telephone, TV, Internet, and news systems.

Philosophy, Section Four (cont.)

I will make my argument that the format of the Production-Distribution System is "playing" a big role in the formation of the "Value System" as a basis, then the "Individual vs. Society" and "Society vs. Society" aspects are a "driving force" in the formation of the Second Level (Chart 16), and this creates the relation for "Economy, Law, Ethics, Morale, Aesthetics, Religions, Art, Politics, Education".

One can read in the definition of "Value System" that it has "Written" rules (example: Constitution, Laws, etc.) and "Unwritten" rules (like how many teaspoons of salt I like in my soup). Because the "Individual" is very important in every "Society" (Society is built from Individuals) I need to spend some time analyzing the individuals a little longer, before I can start making my arguments about "existing, current" and "future, possible" "Value Systems".

I will investigate the attributes of the Individuals more and I must spend some time extending the "**Attributes of Human Beings Theory**" to cover more of the influential and very logical/emotional aspects of the **Individual** that can "play" a major role in the "**Individual vs. Society**" aspect of the "**Creation and Changes**" of the "**Value System**".

We already analyzed the basic very important attributes of human beings, now I will extend those observations with a few new attributes — that is the functions of our "**Self-Realizing, Mirroring, Human Mind**", and it "plays" a very important role in the formation of any "Value System".

Part 2 - Chapter 17
Philosophy

Philosophy, Section Four (cont.)

Let us investigate Chart 43 and Chart 44:

Chart 43.   **The Left and Right Hemisphere of our Level 3 Logical / Reasoning and Level 2 Emotional Mind.**

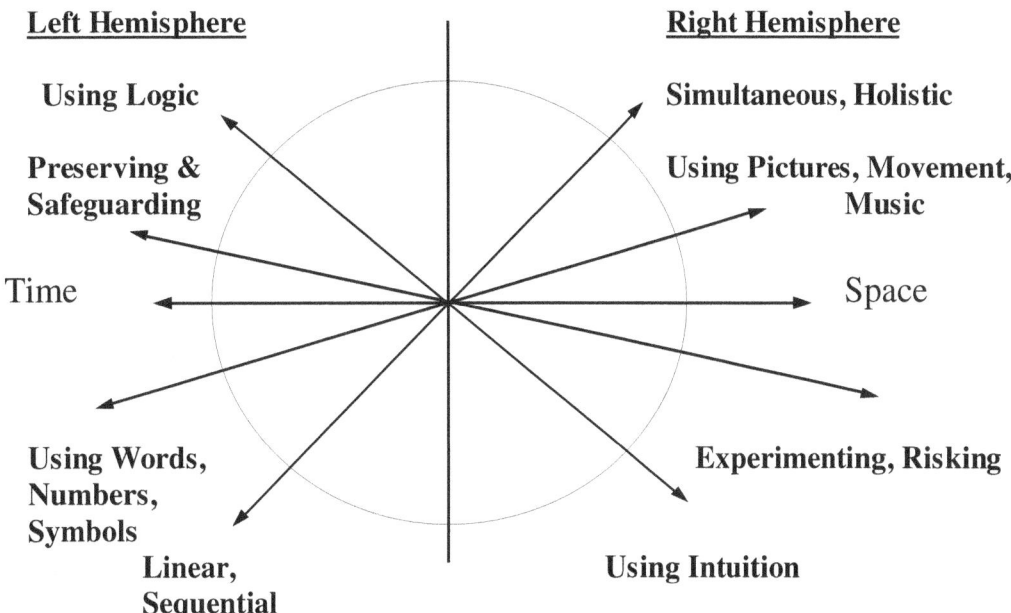

Left Hemisphere

Using Logic

Preserving & Safeguarding

Time

Using Words, Numbers, Symbols

Linear, Sequential

Right Hemisphere

Simultaneous, Holistic

Using Pictures, Movement, Music

Space

Experimenting, Risking

Using Intuition

Chart 44.

Left Hemisphere (Safeguarding Self) **At Work!**	**Right Hemisphere** (Risk-Taking Self)
Makes rules; Avoids wrongness; Is fearful; Avoids risks; Avoids surprises;	Breaks rules; Does not mind being wrong; Sees the fun in things; Takes risks; Likes surprises;
Alert to danger; Is logical; Has to know;	Open to anything; Is Intuitive; Is willing to guess;
Wants everything to fit; Analyzes to pieces; Realistic; Hesitates; Prizes to know; Hates confusion; Intellectualizes; Serious, cautious; Looks at consequences;	Can tolerate loose or no connection; Recognizes patterns; Wishes; Impetuous; Goes cheerfully into unknown; Can tolerate confusion; Imagines; Playful; Ignores or doesn't fear consequences;
Evaluates; Draws on conscious, reassurance;	Feels; Draws on unconscious for reassurance;
Tries to punish mistakes, wrongness of past;	Goes into future without looking back;

273

Part 2 - Chapter 17
Philosophy

Philosophy, Section Four (cont.)

One can see that our "Genetic" makeup can have defining roles in our "Individual Mind's Makeup". It is easy to see from Chart 43 and Chart 44 that the options are simple (actually threefold).

One with a dominating "Right Hemisphere" mind will most likely become an Individual with interest (or professional occupation) in Law, Ethics, Morale, Aesthetics, Religions, Art, Politics, and Education on the Art side, etc.

One with a dominating "Left Hemisphere" mind will most likely become an Individual with interest (or professional occupation) in Sciences (Math, Physics, Chemistry, Computers), Economy and Education in the Science side, etc.

One with a more or less balanced "Left and Right Hemisphere" is the Individual for Philosophy or any other occupation of any side, however, I think strongly that this Individual can't reach the level of outstanding specialization of any of those who are "Only Left" or "Only Right" dominant.

To complete the investigation of the Individual and the extension of the "Attributes of Human Beings Theory", I need to introduce one more observation. I can call the observation of Chart 43 and Chart 44 as a **"Picture"** of the **"Structure of the Minds of the Individuals".** I will draw Chart 45 next to represent the **"Picture"** of the **"Natural (Genetic) and Social Inheritance of the Individuals"**, the concept and definition of the "Knowledge Base" of "Natural and Social Inheritance" in one's "Individual Mind". One needs to understand that all these are very important because in the definition of the "Value System" we already saw a factor of the "Individual vs. Society" to affect the existence and shape of the "Value System".

One more important note, that Chart 43 and Chart 44 describe the three possible makeups of **"All Individuals"**, but Chart 45 will describe only **"One Individual"**, a possible snapshot in a moment in time. One needs to understand that the "Picture" of Chart 45 is different for all individuals and it is constantly changing in time. (Chart 45 is one possible "Snapshot" representation of one possible complete Chart 29 in the "2D (two dimensional) environment".)

Part 2 - Chapter 17
Philosophy

Philosophy, Section Four (cont.)

One can see in Chart 45 that a "**Natural (Genetic) Inheritance**" is coming from the Level 1 Reflexes, that we all (healthy one's) are born with and our Natural (Genetic, defined by our genes!) makeup, like hair, eye color, body shape and size, gender, etc. The "**Social Inheritance**" is coming from our upbringing and experiences, like Home, Extended Family, School, Work Place, Clubs, etc.

Chart 45. **Natural (Genetic) and Social Inheritance of the Individual**

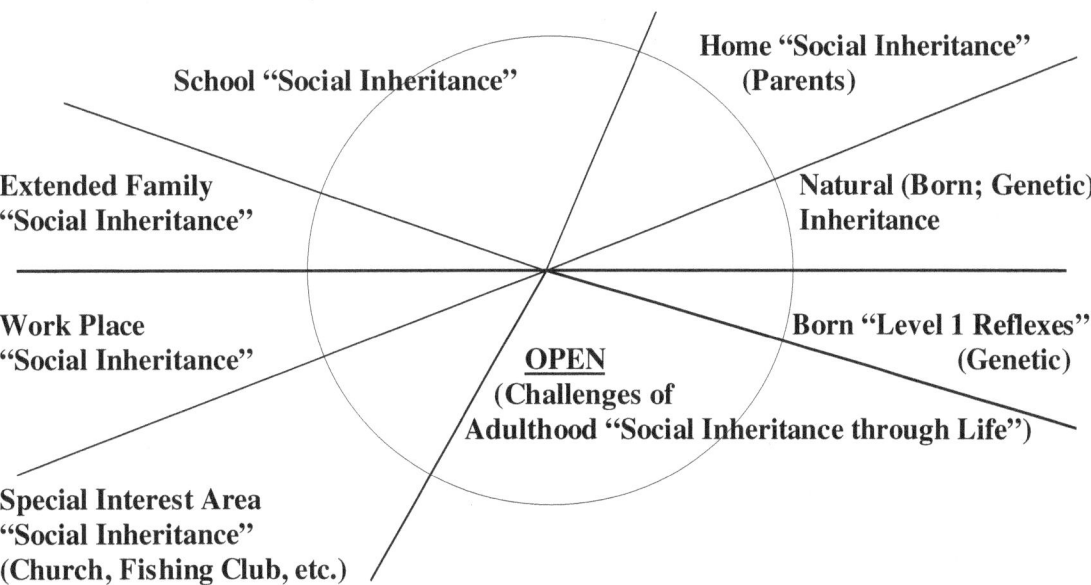

We also have an **OPEN** area, where there is room for our "Values" to develop, change, and form through out our Adult years, in our own ever-changing environment that we are living in. One can see how each Individual can "pick up" the written and unwritten rules and behavior patterns in one's environment growing up. One can ask: Why is it important for an argument about a "Value System"? One needs to realize that the "**screen door**" for an Individual, that I've been referring to so many times, is coming from Chart 45.

The "Natural and Social Inheritance" defines the individual's mind set at every point of time in one's life.

Part 2 - Chapter 17
Philosophy

Philosophy, Section Four (cont.)

The constant interaction between Individual vs. Society form the Individual's and the Society's "Value System". In this process it is very important for a healthy society, to have the healthy (physical, emotional, and mental) upbringing of every individual. This isn't an utopia, it should be a responsibility for the leadership of the society to keep the public health on the highest priority.

For example: A mentally ill individual with no health care coverage can end up in a crime situation, (with a mental problem one can't handle the normal average pressures of life), and can end up killing someone. Now we have a situation where this one individual damaged a whole family of the person who was killed, and we (our society) put this (now) criminal in prison, and are paying for the cost of living for the rest of that criminal's life. Wouldn't it be a better solution to treat the mental illness and avoid all this mess? Maybe with some medication this individual could function as a valuable member of our society. One can see in this example, that early detection can be viewed as an invasion of privacy, so the "emotional freedom" would be violated.

One can ask a question of "what is better for a healthy society, a little short coming on "emotional freedom" (that feels bad) or the unhealthy but free society, with all the mess that those unhealthy individuals create for all?" Where is the freedom in this example? The mentally ill person is destined to end up in a bad situation, if it can't function in a "normal" society.

One also can see that "One Generation" hands down the "Social Inheritance" to the next generation. Here is where another problem begins, because the ever changing, growing environment (**Technical and Human Progress!**) makes some of the rules "**Old**" and obsolete, (in other words, such that didn't serve the progress of society and the next generation anymore!), so changes need to be introduced and this creates a conflict.

Part 2 - Chapter 17
Philosophy

Philosophy, Section Four (cont.)

I will investigate only the top "Nations" here, sticking with my sad, but valid, observation based statement that about 80% of the countries on Earth are not worth living in. (e.g. Those places where they exchange a daughter in marriage for two goats, or for the 120 wives societies also, justified by their religions and traditions, explain strongly the worthlessness of those countries as far as "Value Systems"!)

Those are all unacceptable human "Value Systems", where one human doesn't treat another human as a human being!

The individuals always act based on their "Attributes and Extended Attributes of Human Beings", that determines the needs and comfort level of the individual, and based on the "Natural (Genetic) and Social Inheritance" of the individual. These driving forces govern the actions and reactions of the individuals in the society. Next I will investigate our (USA) current "Value System" (with our existing problems) and a possible "New Value System" with possible solutions.

One needs to realize that not all values need to be changed. Some of the "General" values like "one human should not take a life of another human" can exist indefinitely. The "Interval" and "Specific" values that need to change are actually changing many times through human history. Their nature is to expire and became outdated with the human mind progress in the historical timeline.

I will address the rules, that I think outlived their validity in the "Capitalist" base society, and address the new possible rules that can create a more advanced "McKaneist" society.

The two important parts of the new "Value System" are the "**Artificial, Unequal, Fair Share Distribution System**" and the "**Artificial Fair Value System**". These are corresponding to the structure of Chart 16.

I will investigate many "Existing Values", but not all of them. Those that need to change in my view will have the highest priority in my investigation.

Part 2 - Chapter 17
Philosophy

Philosophy, Section Four (cont.)

This "Philosophical" investigation is the beginning to highlight many of the problems, but as I said earlier, only a society can create a complete "Value System" that is acceptable for all and all are willing to live by it. The following pages of Section Four, will start a process, for humanity, I am positive about that. With this intention to introduce the problems and possible solutions let us move on to analyze the topic of "Value System" in more detail.

In the "civilized" world (like the USA and a few more industrialized nations) today, the "**Value System**" is combined from **three** main "philosophies" (more or less "thought processes"), that is **"Pragmatism"**, **"Existentialism"**, and **"+ 1 Religion"**. (examples for the "+1 Religion": Christianity, Islam, Judaism, etc.)

Most people in these countries don't even know what belief system or "thought processes" they live by! For your introduction, (in the USA) you live by "some mix" of these three, knowing it or not. In short, "Pragmatism" is looking for the "Cash Value" of things, (more car for the money) or you ask a question "Does it work?". "Existentialism" is projecting one's image through one's material existence (net worth). "+ 1 Religion" is by the choice of your parents (at least to start with) to provide the additional moral foundation for one's life. I will analyze these three in a little more detail now.

"**Pragmatism**" with it's "Cash Value"-mania really doesn't give any answers for our morale, health, etc., issues because we can't place a "Cash Value" on certain things in life. (We have this stupid thing in the USA that "everybody has their price", saying that, if the "bucks" are big enough, anybody can be bought). I have news for you, nobody can offer "any sum of money" that can buy me when it comes to my "own health" or the "life of my child", and I can go on and on. Your "Pragmatist", "Cash Value" stuff doesn't work on those things.

"**Existentialism**", that goes into extremes, to show that one can buy the thirteenth castle on top of the previous twelve. We must "**accelerate our emotions**" beyond all beliefs, to show the world that we really are worth more than the others, even if it goes to a wasteful, illogical, unnecessary spending.

Part 2 - Chapter 17
Philosophy

Philosophy, Section Four (cont.)

(Nobody on this planet can prove it, with any logical argument, that one city has to spend many millions on "fireworks" for celebration time to outdo another city, and in the mean time, they have hundreds or thousands of homeless people in their city that could use those resources! Nor can anyone prove that "one needs thirteen castles to function" in life!) The "freedom" argument won't fly here either. (Freedom to buy thirteen castles?!)

We don't have an "Absolute" freedom in our planet as we were told to believe, because our planet Earth is finite, and not infinite in it's existence! "Freedom" is very relative, (that nobody wants to tell you, because you are an emotional, sensitive mind that gets easily depressed!) as I will demonstrate with the following example.

If one places "one ball" on a pool table, that ball will have an "Absolute Freedom" to go and occupy any place on the pool table. If one places "nine balls" on the table then any "one ball" will have only a "Relative Freedom", because it will have eight other places on the table that this "one ball" can't occupy. (because those places are already taken by the others!). One can place so "many balls" on the pool table that none of them will be able to move at all. That will be a "Relative Zero Freedom" of movement. (Looks familiar?, the stop and go traffic on the highways?, how about the situation when we overpopulate our planet? How about 10, 20 billion or more people in the future?) No reason to be depressed, nothing is wrong to understand and live by a "Relative Freedom", because that is reality, we just need to find solutions!

"**+ 1 Religion**" is what we **need** as a component number three, to complement the first two and to **answer those questions** that the first two don't or can't. **Do they, really**?

Of course they don't! They just make the "Value System" more complicated, more confusing. Let me investigate all those few elements that one can find in any of those "Belief Systems" or "Religions".

First: One needs to have a "**Higher Power**" with an "**Institutional Authority**", (Church System) whose words and decisions have a serious saying in the society. (Inquisition can do the trick as far as authority, the opponents are not around anymore — "burned on the stake"!)

Part 2 - Chapter 17
Philosophy

Philosophy, Section Four (cont.)

Second: One needs "**Rituals**" that define appropriate worshipping behavior for the individuals within a society.

Third: One needs to have a "**Speculative Aspect**" of religions, that is explained with the "**Words of the Higher Power**" for the answers for the basic philosophical questions (even if we have no answer for some of them in reality!), such as: Who are we? Why are we? Where are we from?, etc.

Fourth: One needs a "**Tradition**" that is handed down from generation to generation.

Fifth: One needs to have an "**Emotional Belief**", that regardless of reality, carries the individuals in the society.

Sixth: One needs an "**Eternal**" and "**Infinite**" nature of things, to replace our own finite, human, hard to accept, realistic existence.

There you have it! "**Pragmatism, Existentialism, +1 Religion**" — all the answers that one needs in a society! Unfortunately, that isn't as simple as it looks like! This setting doesn't reflect the reality that we are living in. (For example: We have no proof for any "Eternal" existence, but it sure "Feels Good" to think about it, that we can reunite with all our loved ones afterlife somewhere and therefore we don't have to be worried about anything anymore forever!) We have to come down to reality here a "little" (maybe a lot!) and we need to realize that most of these "thought processes" either don't have an answer for our problems in reality, or give us the wrong answer.

One more interesting observation that I have here about the incredible ingenuity of our "Founding Fathers", (USA) who foresaw more than two hundred years ago that our world changes and our (USA) Constitution has to change with it, some point in the future.

Maybe it is the time? They incorporated into the (USA) Constitution, as "Article V", the "vehicle of change" whenever those serious changes become necessary. They did know that the time will come when the original system needs to be changed because of the changing nature of our human existence. (It is amazing to see the "Logical Relativity" in their thinking, more than two hundred years ago!)

Part 2 - Chapter 17
Philosophy

Philosophy, Section Four (cont.)

Now, I will move on to analyze the existing problems, and introduce some examples, (that can be observed by anybody) of our (USA) current system.

"Pragmatism, Existentialism, +1 Religion", that is the **"Value System"** in the USA, and most people will tell me if I bring up any of these subjects or topics that: "It is OK!, What is your problem with this? **If it ain't broken why try to fix it?**".

Let us view Chart 46, to investigate this last statement!

Chart 46.   Status of the USA as of the year 2003:

1. Social Security will be Bankrupt by year 2015, 2020. ("Good luck getting Old", including me"!)
2. 40 million don't have health coverage out of 120 million households (33 1/3%!)
3. 31 million living below the official poverty level (25% of the USA, and they are working!)
4. 7 million word tax code, and we can't collect $280 **B**illion in tax
5. "400" must have $880 **B**illion to "control and play" with the life of 300 million?
6. "400" let our jobs leave the USA , "their wealth" grows faster with cheaper labor!
7. 2 million people in prisons and jails, cost an average of $30 to $40,000 a year each?
8. $40 billion+ for Homeland Defense / year, (Could be cheaper, if other countries do like us instead of hating us! Better foreign policy stupid?!)
9. We are buying friends with $cash$, instead of establishing friendships through common values!
10. We are promoting "World Capitalism", but it doesn't work! (Read arguments in the book!)
11. We are promoting "World Democracy", but it doesn't work! (Read Aristotle 2500 years ago!)

It ain't broken?!, and this is **The best country (USA),** the **only existing superpower**!? (Most other countries aren't even worth the time for the Analysis!) These are just the **"Big Ticket Items"**, and the **"Tip of the Iceberg"** — a lot more can be listed here. (Each of these points can be analyzed in a separate book, that is about the size of these problems!)

Part 2 - Chapter 17
Philosophy

Philosophy, Section Four (cont.)

It ain't broken! You are correct — **It is falling apart!** I have no doubt that this is the best country today, but my questions are: How do we want to lead the world, solve every other country's problem, if we can't fix ours here at home? Does anyone truly think that we can go on like this and our economy won't fall into a complete disaster? How does an unhealthy (no coverage 33%) nation lead a healthy world? How many more hours a day (already 10, 12) do we want to work to pay for the whole world's problems? (Remember?, Tax Payers Money!)

How will we address the population jump to 10 billion by year 2050? We need to remember from history that the economists of the "Capitalist" societies told the so called "Communist" societies that their system doesn't work. The Communists were taking it as a joke, saying: "We are living it every day! What are you talking about? If we can live in it every day, it must work"; then 1990 came along and almost all of them went away overnight!

Now, I am telling you that "Capitalism" is in the same situation today! Don't believe me, just open your eyes, look around in the world, and check out all the problems we have! The next "**should be**" action of humanity is very simple — we need to design and implement the "**Artificial, Unequal, Fair Share Distribution System**" with an "**Artificial Fair Value System**" on "top of it". ("Artificial" means here that it is "human made"!)

The other possibility is to do nothing and the society will change by trials and errors as it did throughout our human history, by the "Directional Theory", (here "Evolution of Society") with all the miserable side effects that we already know from history! (Wars, Revolutions, Chaos, etc.)

Make no mistake about it, the "social structure" in the near future will change with or without our design, either way, because we (humanity) have "Many Big Problems" world wide and our "**Technical and Human Progress**" will force the change! The USA alone could travel on for a longer time, but try to carry the "Many Big Problems" of the 5.7 billion people of this planet on the 300 million people's extremely hard work, and this "will break the camel's back" in the near future!

Part 2 - Chapter 17
Philosophy

Philosophy, Section Four (cont.)

We (humanity) are on planet Earth as of the year 2003, living by "Outdated Value Systems". I am not saying, as one friend of mine always corrects me, that "I <u>think</u> we are living by an Outdated Value System", I am saying, as a <u>"Matter of Fact"</u>, <u>that all 6 billion fellow planet Earth citizens are living every day of our life in the</u> <u>"Outdated Value Systems" of one of the existing nations!</u>

It seems "**very critical**", in some people's eyes a bold, maybe invalid statement, but one can do one's own research and find out that the daily news of the world doesn't tell even half of the story, because of the "political correctness" = "keep the people in a peaceful mood" in the society that they live in! Now, after so much admiration and some criticism of some of my predecessors and their systems, I am ready to take a "Bold Step" forward and introduce my "thought process" for a possible better "**Value System**" that we (humanity) can live by. Interestingly, here in the USA , I think we have all the possible needed conditions (other than maybe the willingness of some people) to make it happen! In this introduction one will find a basic outline and some pointers for the possible new "**Value System**" by Mckaneism. I don't claim that this system is complete and can be implemented tomorrow, because it won't be true, however, with a small imagination everybody can see the points and possible extension of this system to a workable worldwide new "social structure".

The next new McKaneist "**Value System**" could be built from the "**Artificial, Unequal, Fair Share Distribution System**" and the "**Artificial Fair Value System**". Both need to follow the logical rules of "**Time Relativity**" and the "**General**", "**Interval**", and "**Specific**" logical rules. A country such as the USA can set up a new functioning system and **offer the other countries on the planet to merge into it**, as long as they accept the system and the concepts of Chapter 9 and Chapter 16. It can be similar to a concept of the European nations that any other nations can join the European union, if one is at least half as wealthy (ratio of wealth of the nation is 1:2). This system is working well so far for them. That keeps the integrity of the economy of the leading country as not being impacted in a poor way.

Philosophy, Section Four (cont.)

Before I detail my McKaneist "Value System" I have to introduce a new concept. We need a "**Free Institute of Philosophy**" (more about it in Section Five) that can work on the detailed design of the new "Value System", yet has no "Power" (Legislative, Executive or Judicial!) and the "Mission should be the creation of a "Value System" that supports the interest of Humanity"! The "Value System" could include the basic economic structure, and the Constitutional and other Legal Laws. (Based on the Level 3 Logical/Reasoning and Level 2 Emotional design!) The System can be presented to an "**Institute of Execution**". (Legislative and Executive merged, with Executive function only, that doesn't create rules but only Approves or Disapproves the "Value System" and throws it back to rework if it is Disapproved!) When Approved, they present the final for the people to Approve. If 90% pass the "Value System" as presented, the system can become the new "Value System" of the land (Planet Earth!). The third component, the "**Judicial Institute**", can function close to what it does today by overseeing, interpretation, etc. (That can sum up the new three balanced branches!) **The Interest of Humanity has the only POWER!**

The "**Free Institute of Philosophy**" should try to design a simple, easy to understand, working, fair system for all. (Mission for them!) It can use for an example, that we have here in the (USA), a logically designed system, that works with minimal supervision. That is the system that I will use as a prototype for my logical system argument. That is our "Traffic Control" system. About 40 to 80 signs control about 240 million passenger cars, a large number of commercial traffic, 240+ million drivers a day, on millions of miles of road. The system is logical, fair, and working!

(No one can invent a more fair system than a "four way stop"! Coming and going in the sequence of arrival!) A few important points here: <u>Point one</u>, we (humanity) can design working, logical, artificial systems, (we already have the traffic control system, for one example). <u>Point two</u>, in this system we don't use the "Emotional Freedom" concept at all, (one may feel emotionally free driving on what seems like an endless millions of miles of roads), but instead this system is designed by the logical restrictions of you "**can-do**" this and that, and everything else is "**no-no**".

Part 2 - Chapter 17
Philosophy

Philosophy, Section Four (cont.)

That leads me to a statement that in system design we need to focus on the "straight forward can do's and don'ts" instead of having our current system, where you say that you are "free to do" everything, but not this, not that, not this … I experienced this approach in computer science, where some people design systems for the exceptions rather than for the main business rules! Guess which method works best?!

When our founding fathers designed our constitution (USA), life was much more simple, so they designed it for a "Freedom Concept", because there was very little "**no-no**" in our daily life, and a lot of "**can-do**". The real freedom and our "emotional" freedom was very close. Today, as life gets more complicated and our population grows to around 300 million people in the USA, we are having more "**no-no's**" than "**can-do's**" in our every day life. Anywhere one turns, another law "pops up" to govern our life. (Congress votes every day for new laws of the land!) Instead, to be "free" we are losing our freedom within the huge amount of laws that no one can follow or understand, because it is so complicated. (e.g. Tax law?!) We need to start with the basic Level 3 Logical/Reasoning rule set, then we can build our Level 2 Emotional freedom on the top of it.

The Level 3 Logical/Reasoning rule set is the more difficult to design; Level 2 Emotional freedom types are the easier ones. Most of our "Emotional, Free Activities" are relatively simple as going to the ball game, going to the mall to go shopping, working out in the health club, going out fishing, going to the amusement park, etc. — these do not need any complicated laws to govern or coordinate their relatively simple activities.

With a well designed system, we can have a smaller, much less expensive government, that can be a "**Guardian**" of the system like "Highway Patrols" for the "Traffic Control System". We can teach to our children in our schools what are the rules of living, and what is the expectation of the family, extended family, neighborhoods — and the society from the young ones, as they grow up. (Social Inheritance!) That could lead to a working logical and emotional society, where everybody knows the rules, beforehand, and not finding out after one's broken it. That is basically the new concept in short, now we can see the basic outline to start with.

Part 2 - Chapter 17
Philosophy

Philosophy, Section Four (cont.)

The **First Component** of the new McKaneist "**Value System**" is the "**Artificial, Unequal, Fair Share Distribution System**". This system was detailed in this book in Chapter 14, modeling the percentage (%) based distribution system. If we add a few more following simple rules, that system can be used as a base for our new Production-Distribution System.

The **First** component of the new McKaneist "**Value System**", the "**Artificial Fair Share Distribution System**", will serve the rules of "**Time Relativity**" and "**Interval**" rules, and these can be revisited and examined by the "**Free Institute of Philosophy**" for Change, Modification, Enhancement, etc., from time to time (like once a year or once every five years for example).

These rules are "**General**" rules, so they will change very slowly, forming the base for the economy of the world society to function properly. They also are Level 3 Logical/Reasoning rules with very little Emotional impact attached to them, so they can serve the "Discipline" and "Fairness" in the society. They also keep the Unequal side of the society to conform to the unequal Attributes of the Individual Human Beings.

We all have our interest areas and there are countless emotional choices as "freedom" for us. Unfortunately for us (in the USA) the trend is that more and more nations want to take a bigger and bigger piece of the distribution of the products of our planet. (Growing trade deficit with many nations every month!) This will lead our standard of living (USA) to decline and the discovery of the people that we need a new and better structure for our society, like an "Artificial, Unequal, Fair Share Distribution System" for starters, then the "Artificial, Unequal, Need Based Free Distribution System" of McKaneism, to lead humanity into a better realistic future. Both of these last two new distribution systems are Level 3 Logical / Level 2 Emotional based.

In the next pages, I will outline the basics of the "Artificial, Unequal, Fair Share Distribution System" in detail, and in a few short sentences, I will define the major differences with the "Artificial, Unequal, Need Based Free Distribution System".

Part 2 - Chapter 17
Philosophy

Philosophy, Section Four (cont.)

One needs to understand that these systems are only theoretical, as of today, but the "Artificial, Unequal, Fair Share Distribution System" can be implemented tomorrow, because only the readiness and willingness of human beings is the dependency for that.

The "**Big Difference**" between the "Artificial, Unequal, Fair Share Distribution System" and the "Artificial, Unequal, Need Based Free Distribution System" is that in the "Fair Share..." **people are involved** in the production while in the "Need Based Free..." the **production is automated** (in that level of human advancement, robotics in production replace humans) and maybe a few people supervise the automated production. In both systems the distribution is "Unequal" and "Fair" or "Need Based".

One needs to understand that the "Artificial, Unequal, Need Based Free Distribution System" will be a perfect copy of the "Natural Distribution System" (Chart 47) with the only difference (philosophically) that the "Natural..." is **nature based** while the "Need Based Free..." is **human made artificial**. (One doesn't find the "Need Based Free Distribution System" in Chart 17, (remember, Part 1 for my "beginners"), but I will create the updated Chart 47 for my "advanced" readers on the next page.

One can see easily that with the "Artificial, Unequal, Need Based Free Distribution System" we achieve the same result as our natural world. Like the best example comes from the "Natural Kingdom" where every participant takes a **need based share distribution** out of the **natural production** environment with zero pollution and one hundred percent recycling!

One also can see it easily that the Second level (Chart 16) social rules will be different in both systems. It would be speculative and science fiction for me if I tried to describe the "social structure" of the "Artificial, Unequal, Need Based Free Distribution System", because just to understand a structure where people don't ever have to work (unless one chooses to) opens up an unimaginable "**Level 3 designed, Logic based, Realistic, Free**" society.

Part 2 - Chapter 17
Philosophy

Philosophy, Section Four (cont.)

I wish I could live in one like that but unfortunately, for me, I won't be around long enough to live to see it. No hard feelings, our kids will be there, so all this writing is worth every second of the effort to guide their world into a better existence! However, I strongly know that the "Artificial, Unequal, Fair Share Distribution System" isn't far away.

Here is the extended (Chart 17), the complete Chart 47:

Chart 47

Unequal Unfair	Unequal Fair	Equal Unfair	Equal Fair
2. "Primitive Tribal system"	1. "Natural Distribution system"		6. "Socialist, Communist system"
3. "Slave State societies system"	**7. McKaneist, "Artificial, Unequal Fair Share Distribution System"**		
4. "Feudal State societies system"	**8. McKaneist, "Artificial, Unequal Need Based Free Distribution System"**		
5. "Capitalist societies system"			

Humanity will experience a level of problems in practical life with "Capitalism", to conclude (if my arguments in this book didn't convince one yet, reality will in the near future) that the "Social Structure" changes are inevitable and necessary for Humanity.

Part 2 - Chapter 17
Philosophy

Philosophy, Section Four (cont.)

Many times I referred to the name of this publication as the "Foundation of Humanity" — meaning it is the foundation for the next available, makes sense "social structure", however, **I never intended to describe how the society "Must!? be Structured in detail" or how humanity "Must!? Live". That would be a mistake from me to try to "Dictate!" for 6 billion people their future without their free choices**.

Having said that, and with those thoughts in mind, I will venture into the next few pages to give some "Pointers" for a "One possible structure" of the "First Level" and "Second Level" (Chart 16) of the society, based on the "Unequal", "Fair Share Distribution System".

My observation is that the most simple system works best in nature, and that can apply to humanity also. The base for the "Unequal", "Fair Share Distribution System" is the percentage (%) based production-distribution model, that I outlined with an example in Chapter 14. That can be the "First Level" (Chart 16) for the new "social structure".

Few more additional thought to this set up. Today (2003) in the USA, the per capita income is around $21,000. The president (of the USA), the last time I checked, made around $300,000. That is about a 1 : 15 ratio in the income level of per capita to president. We can use this model to start with in our new percentage (%) based company structure, where the CEO can't have more than 15 times the pay of the average worker (by the way, interestingly, this ratio was existing in the 50's of the USA, a ratio around 15 to 20 times, CEO vs. average worker).

That was a real picture in the USA before the "Super Greedy" rich group of people developed in this nation to destroy the world for all the rest. (Today's CEO is paid $10-15 million income within a bankrupt company, where the average "Joe" makes $30-40 thousand a year! That is a ratio of 250 to 500 times! Another recipe for disaster on a national scale! Pick up this issue as a national security interest! If your economy falls apart, your whole nation could fall apart!)

Part 2 - Chapter 17
Philosophy

Philosophy, Section Four (cont.)

As far as the owners, in the new "Fair Share Distribution System", they can set their percentage (%) for starting the company (they bring the capital), however, if they are not realistic and fair, the company's venture will not work (it won't even start!). It will take some time before people start to understand the new concept of "Fairness" in the distribution system (especially the top "dogs" who have it all today!), so the society has to set a maximum limit of wealth. (by law!)

We need to **Limit the "wealth" (net worth) to around $100 million for one Individual**, (or any other number that seems "**Fair**" for all!?) and that will help recycle a huge amount of "currently dead capital" into new upcoming small entrepreneurs' ventures. (Preferably in the USA first! We need to "get well" in our own USA first before we can help out the whole world!) I don't worry about how much one individual makes or spends in a year (One can make billions and spend billions, but we need to understand that at the end of the tax year, one can't have more than $100 million maximum wealth (net worth!). More wealth than that could give one too much control and power in the society over politics, markets, economy, monopoly status in business, etc.

One can buy many local elections, lobbyists, members of the parties, etc. It would be the same scenario that we had with the power of the King at the French revolution in 1789, or the power of the English King at our Constitution time in the USA. Too much power in one or in a few hands can destabilize the society and work against the Fairness!. Also, we need to take "cheating" on this limit very seriously. The best punishment is to take all the wealth from those who cheat on this maximum rule. (Like three strikes, one is out! If $100 million is not enough, let them try on nothing, starting from zero!)

I hope that one already understands from the previous arguments, that in a "closed system" (Planet Earth), when one has way too much, that means that many others can have nothing. The practical limit of the Wealth of the individual can be set somewhere around 100 million dollars (in 2003 dollars) for starters.

Part 2 - Chapter 17
Philosophy

Philosophy, Section Four (cont.)

Understand here that we keep the "Unequal" nature of capitalism, and we also keep the high achievers with the possibility to supercede the average person, but with this concept, or later hopefully with an enforced law, we can throw out the "emotional Greed" and the "huge resources" in one's hand to "enslave the society" through financing the whole election process and enforcing one's (or "small group" in practice today) view in the election results.

Society can adjust this top wealth limit any time based on the "fairness" concept in any given time. One can notice one more thing, that this limitation doesn't limit an individual's freedom, as in "how many" and "what kind" of assets one can have. It only limits the final top size of it. One can see the flexibility of this new system.

Remember from earlier, that in our USA we have 400 people with $880 billion of wealth. I am not proposing that we take their wealth away tomorrow, however, one can imagine the huge economic impact if those 400 would have $40 billion of wealth (400 x $100 million) and the remaining $840 billion can be recycled to our society as venture capital to start new companies in the percentage (%) based structure. If the USA did this, all the other nations would go crazy from the economic impact that we could deliver on this planet (The Superpower (USA) will be a Super-Superpower!).

After all, some of you "very rich" individuals need to understand that having it in 12 to 13 different 30 bedroom castles and 10 to 15 half million dollar's worth of cars in your garage is not necessary, even for a "rich lifestyle" — only to "brag about it" to each other. In the mean time, all of you live in a society where you are "prisoners" of your own castle, because the outside world is falling a part around you in the rate of the speed of light. I personally think that nobody supports "Fairness" in any system; by taking so much out of it, that the system starts collapsing.

The next big move can be to change the Tax system. In a "Fair Share Distribution System" based society, one needs to have a fair tax system.

Forget the 7 million+ word tax code (year 2003 USA) and introduce something simple; a better working tax system.

Part 2 - Chapter 17
Philosophy

Philosophy, Section Four (cont.)

A few pointers here. **One simple percentage (%) rate for all incomes**. It is just as wrong to charge 28% or 36% for the "rich folks" and 10% for the "poor folks" as it is wrong that someone can have billions of dollars of wealth and the others are forced to sleep under a bridge, flat broke.

There should not be any corporate tax or any tax deduction for anything (that is social engineering! Remember the competition between the "Holders of the Capital" and "Big Brother"?) The companies can operate on the percentage (%) based distribution system and spend some of their profit on the basic cost of their business. The rest of the profit could be distributed between all the "players" in the company (Chapter 14) based on their percentage (%) structure and taxed at the individual level. One can see that this is a simple system. If one thinks that "bad people" can spend company money for their private use and cheat the system, think about it — it doesn't matter because where one owner can acquire one's wealth, whether it is corporate or personal (from investment), it can't be more than $100 million tops.

One can see that it is not easy or possible to cheat the "Fair Share Distribution System". If one tries to underpay one's employees (smaller than a fair percentage %), or tries to create one's wealth to exploit the company's wealth, that one will be eaten alive by one's business competition. Just remember, I never had any intention to throw away the "Free Market"! In the "Fair Share Distribution System" the "Free Market" is even functioning better, because everyday people have **more interest** working within the "Free Market" economy. One really shouldn't care that one will have three castles, or that one will have 2 company jet planes, or another one will own 5% of 200 companies. The distribution has to stay flexible and "Unequal", because that part of capitalism works, only the "Unfair" part needs to be changed to "Fair". I recommend setting up a few models, then one can see for oneself how this system really can work and make more sense in reality than the current one.

Part 2 - Chapter 17
Philosophy

Philosophy, Section Four (cont.)

For the "super rich one's", this one needs to see that this system is coming, one way or the other (I hope one sees this reading my book so far), so one "super rich one" can be a celebrated part of the solution or a "sorry loser" if one wants to stay as part of the problem!

The "**Tax Law**" could be simplified easily to replace it with a fair and easy set of rules (% Tax). One more example here for the "**One simple percentage (%) rate for all income**" — use an example of **17%** (that is a theoretical percentage (%) for this example and not necessarily a recommendation. A simple form, that asks what was one's income and lists the values for "Net Worth" — the check to see if it is close to the $100 million maximum limit. It could be easy for a (**simplified!**) tax institute to track those, because their number is not too big. The rest pay a simple percentage (%) on the yearly income. No income tax for companies, No tax deduction to anyone for anything — Individual or Business, No inheritance tax, No sales tax. Same percentage (%) based tax can work for State and Local Governments. We need to keep the Free Market, Stock Market, Commodity Market, etc. — its just time for a complete overhaul).

Our economists (in the USA) must come up with a feasible percentage (%) rate based on their economic modeling, not a philosopher. I can recommend a logical new structure, but they must figure out the possible implementation (that is their knowledge!). If one makes $10 million in a year (remember, it fits in the $100 million limit on assets!) that one could pay 17% or $1.7 million in tax. If one makes $10,000 a year that one could pay $1,700 in tax. (this example, based on the previous 17% theoretical percentage example)

It looks very "Fair" to me. The "Taxation" should cover a few simple "Need Based" things instead of our current "Social Engineering".

The following items must be the starter items on the list:

1. Defense of the nation.

    1a.   Armed forces, for our national protection (self defense!)

    1b.   External / Internal Security forces.

    1c.   Fire Protection forces.

    1d.   Internal Security (homeland security, like airport security etc.)

Part 2 - Chapter 17
Philosophy

Philosophy, Section Four (cont.)

2. Health Care for all citizens.

3. Social Security for the elderly.

(One can see that in the future, people on the percentage (%) based pay system will have a greater opportunity to become financially independent. That can be the starting point to phase out Social Security programs in the future.) Of course, these are just the core recommendations, we need to revisit these issues at "design" time.

Next, every organization has to incorporate for profit. Remember, no corporate tax, so if one company distributes all profit via percentage (%) to all stakeholders and employees, then no money is left to argue about. We need to get away from these "Privileged Non-Profit" organizations in a "Fair" system.

I was reading a statistic about five years ago, that the number of government employees superseded the number of industrial workers. Lately it seems like it is a "good choice" to work for one of our governments rather than work as a productive worker in our society. One can get insulted, when I say categorically, that most of the employees of the governments are not too productive.

One probably guessed by now that I will recommend that they take the pay to the governments (Federal, State, County, City) to the level of the per capita income of the society. That is about $21,000 in the year 2003. One can see that we will accomplish two major things with this change.

One, that those who use the government job to get rich will quit in the "Fair Share Distribution System" driven society, and that is a good thing for all of us. **Two**, that instead of voting for their own raise, the raise of the government employees will be driven by performance. If they implement good suggestions and rules to accelerate the "Fair Share Distribution System", their income will grow automatically every year. If the per capita income doubles, their paycheck would double. Sounds very "Fair" to me.

One can recognize by now that the major changes in the "First Level" (Chart 16) will trigger major changes in the "Second Level" (Chart 16).

Part 2 - Chapter 17
Philosophy

Philosophy, Section Four (cont.)

I pointed out some possible changes in the "First Level" (Chart 16) system (as pointers only), but I will be more than happy to sit down with anyone (individually or a group as a company) who is interested and draw up an extended, doable blue print for the implementation of the "Fair Share Distribution System".

I will analyze some of the hypothetical possibilities changing the "Second Level" (Chart 16) later and it is there that I will address the possible "value system" of the society and how the "First Level" influences the relations between the values that human societies live by. One will see that these changes could be very unpopular for the current leading group, because the "Fairness" of the society takes away some of their "Unfair" privileges.

One needs to understand that it isn't me who is trying to take their privileges away, it is the ever moving society with the drive for "fairness" in the society. The constant change of the direction of "Equilibrium", that brings all these changes to society throughout history.

We (humanity) have to depart from the emotion based system, and we (humanity) need to design a Level 3 Logical based system with the properly disciplined emotional harmony merged into it (Chart 11), with a new concept of the new individual development via teaching and schooling system.

The next part of our observations and investigation will lead us to a preparedness for a part of section four, that of an introduction to the **Second Component** of the new McKaneist "**Value System**" that is the "**Artificial Fair Value System**". This is the part where I will analyze the Second Level (Chart 16) "value system" of the societies (especially the USA).

The **Second Component** of the new McKaneist "**Value System**" is the "**Artificial Fair Value System**". Defining this is more difficult than the First component. This is the part where we encounter our own personal Individual "freedom", emotional or real in the Society. These rules will also serve the rules of "**Time Relativity**" and "**Interval**" rules, and these too can be revisited and examined by the "**Free Institute of Philosophy**" and could be Changed just like the First component rules.

Part 2 - Chapter 17
Philosophy

Philosophy, Section Four (cont.)

We need to focus on that we don't create more than necessary "**General**" rules and leave a lot of openings for the "**Specific**" rules that represent our own Individual freedom. We need to construct these rules from the viewpoint of the Interest of Humanity!

In my views and recommendations, the "<u>**Highest Values of Humanity**</u>" are for an "**Artificial Fair Value System**" that has:

1. "**Self Preservation of Humanity (and each Individual's within Humanity)**"
2. "**Health of the Individuals**"
3. "**Our Children's Survival and Health**"
4. "**Our Own Individual Time in Our Own Life (Every second of our lifetime)**"

The priorities of these are just as important in this order as I outlined them!

One may be surprised not to find here: "Wealth", "Castle", "Sports Car", "Yacht", "Power?", "Fame", "Religion", "High Social Position?".

On the other hand, one can find the "Care and Respect for the Existence of Humanity", "The Importance of our own Individual Health", "The Love and Care for our Healthy Children (of course within a family unit!)" and the "Individual Time of our Life". These are the most important values (not exclusively) that we should build a "Social Structure" around. One needs to understand that the priority is extremely important.

If the "Existence of Humanity" disappears, the rest is very much unimportant. (One can't have rules and a "Value System" if one doesn't exist!) Without good health, one won't be able to take care of one's children. (If one can't support oneself, one isn't likely to be able to support one's children!) And of course, our own time is one of our most important assets.

(For those of you who will get the "Real Thing" in Heaven, I just say that I want to have my "Little Piece of Heaven" right here on Earth, and I will work on the rest when I get there!)

The "<u>**First**</u>" and most important value in my recommended "**Artificial Fair Value System**" is the "**Self Preservation of Humanity (and each Individual within Humanity)**".

Part 2 - Chapter 17
Philosophy

Philosophy, Section Four (cont.)

We need to understand that we have forces around us like "Earthquakes", "Mega Tsunamis", "Mega Volcanoes", or the possible force of an extinction level of impact of a "Meteorite Strike" to our planet. These are forces that we never encountered in our written human history, and the size of those forces are so great that we have very little or no protection or action against them.

The next interesting question is "What kind of idiots?" would build their home on a 6,000° F hot, melted, rock and iron foundation? Let me give you a short answer. We (humanity) are those idiots! The center of our planet Earth is that hot, melted, rock and iron core. The balance of physics and chemistry let us believe in the false impression that everything is OK. We are living in a very unstable environment that can change at any time to the extent that it can become uninhabitable for humanity. (We see occasionally what is inside, when volcanoes erupt!)

Remember the "Directional Theory", if the change is fast, it usually is the living entities (in this case humans, animals, plants) that can't adjust to the changes. They are extinct. (It happened to T-rex. Why do you think we found all the bones? To look at them and say "gee"? No, I think it was to give us fair warning that it is time for change.) The point here is that we (humanity) should focus all our effort and finances to overcome and master these forces that threaten our survival as an intelligent species.

One can easily see that we can't oppose those forces, but we can avoid them and "move out" our existence to space. If one's "home" is circling around Earth, then Tornadoes, Flooding, Earthquakes, Volcanoes, and Meteor Strikes can't touch you and won't harm you! Sounds impossible? Flying was once-upon-a-time too. To preserve "each Individual within Humanity", we also have to stop overpopulation and mass starvation.

The "<u>Second</u>" important value of the "**Health of the Individuals**" should be self evident, however, we (humanity), especially in the USA, are very confused about it. I addressed this before, that our Democrats think "Health is a Right" for the individual, (coming from a religious fate based belief that God gave us the right to Health! No he didn't!).

Philosophy, Section Four (cont.)

Our Republicans believe that "Health is a Commodity", that you can buy through health insurance. (the good old Pragmatism, that puts a "Cash Value" on one's "heart or lungs"!) Don't! It is one's attribute and you can't live without it!

Of course they are both wrong — "Health is a Human Attribute"! One needs to realize that some "lucky individuals" can go through life with little or no problems, but some are "not that lucky individual" (remember "Genetic Inheritance"!) and need a lot of heath care. One individual is never able to afford the cost of their own health care. For a "Healthy Nation", one can see that we need a "**National Health Program**", that covers everybody. Everybody pays in, and whoever needs it is the one who uses it! I hope one recognized that I didn't say "National Health Insurance Program". Insurance doesn't work and I will address that too. It should be a non-bureaucratic (computer based) "**One Payer Trust Fund**". The reason Health Insurance doesn't work is because insurance companies are running to make a profit in the free market business environment.

No profit and the company can't exist. So they spend unbelievable time and effort to calculate who they don't want to insure, because of the high risk for their profit line — and 40 million without health care in the USA, looks like great insurance risk management (congratulations, you did it!) and a very unhealthy society!

The "**Third**" important value is **"Our Children's Survival and Health".** One needs to understand that for our children (that is the continuation of the human race) we need to take care of the **First** and **Second** important issues, then we need to investigate the "Personal Development" of our children. Personal Development, based on the "Knowledge Chart Diagram" concept, is more than education. It builds the "Value System" and "Personality" with harmony of the "Knowledge Based" education.

We need to reform the "school system" to teach more of a realistic science (evolution maybe?), philosophy, USA and world history, and set their "Value System" from their early age, instead of teaching the "Emotional Freedom" idea. (Then we are surprised: Why don't they obey the law?

Part 2 - Chapter 17
Philosophy

Philosophy, Section Four (cont.)

It is simple: Obeying the Law requires a disciplined, controlled behavior, not (as many think) unlimited "Emotional Freedom"! An "Emotionally Free Mind" is undisciplined! A "Logical Free Mind" is trained and disciplined!)

For example: when I was growing up in the "communist" society, they didn't have any law like we have in the USA about not selling alcohol to under age (18) kids. Many times I was asked by my parents to run to the small corner store and bring home a few bottles of beer when we had a surprise visit from relatives. I was age 12-14 and it never ever occurred to me to buy one bottle for myself or open up a bottle and drink it.

Not because I was an "angel" kid, but because the "Value System" and "Social Expectation" of my parents, relatives, and the society (Social Inheritance) simply didn't approve it. Growing up they reinforced their disapproval many times verbally, so one kid gets conditioned many times not to do the wrong or socially unacceptable thing. Here in the USA, our kids hear twenty times a day that "you are free", and from the age of 15-16, many times, the friendly police man becomes an everyday visitor to the family to (we have a system where police must help to raise a kid!?) tell the kids what are they free to do and what not to do!

The "**Fourth**" important value is owning **"Our Own Individual Time in Our Own Life (Every second of our lifetime)".** We call it today, in (USA), financial independence, where most of the time one's money works and the individual doesn't have to. With the reduction of the number of the people on Earth (voluntary population control), and the use and advancement of robotics, we can achieve, that we can own our own time.

This is an introduction of a "**New Value**" in this country (USA), where we are currently willing to work countless hours for more money. The concept is coming from Existentialism, that we are what we project out. Having more makes us more! Unfortunately, this isn't the truth. We give up our time happily (That is where religion jumps in!) because we will have plenty of time in Heaven to enjoy. How about just "hypothetically speaking" we don't have Heaven. (I didn't see any home videos from there yet!) Then we wasted all our existence here and that is it!

Part 2 - Chapter 17
Philosophy

Philosophy, Section Four (cont.)

When life ends, it ends! (Hard to think realistically!?) One needs to understand that it is a "good, healthy, and smart goal" to use our life here on planet Earth wisely, and enjoy as much free time (Time that we own and we control!) as we can.

Instead of working 10, 12, 14 or more hours to buy more stuff, to fill up our homes with mostly useless "things" (with no real realistic value) and "feel good" about it! (We just can't stop our undisciplined mind from going shopping!) Have any of you ever wondered (I did!) about when you are moving from one of your homes to the next, how much stuff we've collected between the two moves and how much stuff we need to move, that we sometimes don't even remember when and why we bought it?

Those are the "things or stuff" that one exchanged one's free time of one's life for. Working for more "toys" and a lot less to enjoy and live "Free" in this life. Most European companies give 4-5 weeks vacation time for their employees, (make no mistake about it Europe doesn't do any better than any other continent when it comes to non-working "Value Systems", however, they are ahead of us (USA) when it comes to free time!) and we have mostly 2 weeks here in the USA. Some European companies try to work on the 36 hour work week and we here in the USA do 50-60 hours or more in a week.

We are proud to be the "most productive" workforce in the world. "Most productive", that is true!, so our goofy political leaders can give away a big part of our money to buy friendship all around the World and we, consequently, have no "free" life! Check out $25+ billion or more foreign aid per year, or the IMF (International Monetary Fund) to bail out loser nations, or to buy their friendship! $400 billion military per year, and at least 25% of $100 billion is for those who are stationed all around the world. "Little! Trade Deficit" every month to make our trading partners "feel good" and friendly to us, etc. I think if I do serious research I may be able to compile a list of items that currently cost around $240-360 billion a year.

That is about $2,000-$3,000 per household per year which is about $167-250 per month. (If one takes $100 a month and puts it in an 8% return on investment for 40 years, that one can retire with around $350,000!)

300

Part 2 - Chapter 17
Philosophy

Philosophy, Section Four (cont.)

What I am saying is that we all (120 million households) can retire financial independent with a $1/2 million in our pocket just on our "Generosity" to the world! That is how much our "proud productivity" is worth!

One can wonder, are we here in the USA too greedy to make money in every minute in our life, to buy all the stuff that is in the market, or are we so "Emotionally Brainwashed" that "we can't feel good" anymore without the constant shopping to feel "in control" in our life, that we "already lost" a long time ago? We need to realize that there is a lot more to life than money, money, and more money!

One more scary statistic that I hope will make you think and to drive my argument home. An average individual, with 75 years lifetime, spends one's life as follows in Chart 48:

Chart 48.

Sleeps	25	years (average 8 hours a day)
Works	25	years (that is only 8 hours a day, not the "proud" 10 to 14)
Food time	10-11	years (preparation and eating)
Bathroom time	5-6	years
Traffic time	3-6	years
TV news time	1-3	years
Time with kids	5-6	years

"Free Time", Vacations
and one's choosing,
for a whole life time" 1.4 - 2.8 years (taking 1 or 2 weeks vacation off per year!)

That is about 75 years (don't believe me, do your own time calculation!).

These were only the most important values in my view, of course, we need to keep the "Good and up-to-date part of our Constitution", and some of the working laws. (like the Traffic Control System!)   A few more possible additions to the **"Artificial Fair Value System"**:

The **"Gun Laws"** can be very simple, we don't need hundreds of pages of regulations and laws! The Gun problem is a behavioral problem!

Philosophy, Section Four (cont.)

We don't want to address it, because it can look bad for our "Emotional" freedom! Controlling ones behavior???, where is the freedom then? Well, if you have a Gun in your hand you don't have a "Freedom", you must behave by the discipline of gun handling and the rules of gun usage! You want to have your "unlimited freedom?"— don't touch a gun!

The two simple rules: 1. Whatever "Gun" one owns, one can use it for "Self Defense" (to save one's own life), or 2. "Legally Hunting" for the meat of the wild animals. I don't care if one wants to hunt a deer with a "tank bullet", as long as one takes no more deer than the number of the legal limit and doesn't shoot anything else. (good luck to have that trophy!) As far as 120 million households in the USA, having one long gun and two handguns on average, I like it! (Amendments Article II) It is the genius of our founding fathers to realize it is impossible to conquer a country and take away it's freedom where every household is armed and ready to defend it! (120 million people Armed Forces! Can't beat that! **My Advise**: Don't try to disarm it, the security of the nation is most important! Just teach gun owners for these two simple rules!)

The influence of "**Free Religions**" should be kept out of the Government, State, County, and City affairs, as our founding fathers designed it in the Constitution originally.

Separation of Church and State! We are experiencing more and more Church influence in our everyday life, because the "Non-Profit" (that should change "For Profit", like any other business) nature of the religious organizations is getting too rich and they use their money to influence the State affairs! (One needs to remember that Jesus didn't travel with a "Cadillac Limo" or "Own a Church twice the size of the State Capitol", so keep it that way, go for the "**fate**", not for the money!)

One needs to think of the "**Third**" important value of **"Our Children's Survival and Health".** That can lead to an early detection of the non-functioning individuals with "**Physical or Mental Health problems**", and that can change the "**Criminal System**" to a "**Medical Early Detection and Correctional System**". The individuals can get early problem detection and treatment, instead of waiting for years until they have grown up as an "ill" adult and will commit a crime.

Part 2 - Chapter 17
Philosophy

Philosophy, Section Four (cont.)

Most individuals can be cured if detected early, with Medications or Rehabilitation and Work programs, where the whole society benefits from getting a productive citizen with one's health restored, instead of an "adult damaged one" that gets jailed or locked in prison for a long time (maybe life) and the society then can pay for it.

These were a few starting pointers of the possible new "**Value System**". One needs to realize that building a "Logical Value System" will take some getting use to. Our human survival, the proposed rule number one, should be more in focus.

A few more comments about values. I have to say that I am really tired of listening to all those "politically correct politicians", who try to create fame and fortune for themselves, constantly bashing Lenin for his communism. As I analyzed before, it is true that his communist system didn't work but he was way ahead of today's "brain dead politicians" by recognizing (following "Aristotle", and our "Founding Fathers" (USA) in the line) that society is an artificial system, and it has to be designed (remember Constitution!), not simply voting "yea" and "nay" for money spending or meaningless rules as they do in our current society. (For example: "Vehicle emissions test" for small people is for the "free bucks" for the government, and the commercial vehicles drive like "smoking chimneys" with no control over them. Of course, they have a "million bucks" lobbying group that a small individual doesn't have!)

Next is that one needs to realize that money isn't a value system, and money can't buy a value system. Money has a value, as a tool of our product's exchange, but as I argued before, there are many things in life that money can't buy. The reason that I am investigating this subject is because we have a few very bad trends in our society (USA). I have no problems with young musicians who are talented and make a lot of money fast, using the CD business. However, I have a big problem where our young ones are looking at these musicians as idols. Some of these "dudes" are very questionable characters; some of their messages through their music is incorrect, and some of their behavior is simply unacceptable in a civilized society.

Part 2 - Chapter 17
Philosophy

Philosophy, Section Four (cont.)

This isn't an issue of free music or freedom of expression, it is an issue where the "rules of aesthetics" are replaced by roothless marketing and "money-making-mania", concluding that if it makes a lot of money, it must be good stuff (typical capitalism).

One needs to realize that aesthetics, as a science of the art, may need to play some function here from time to time if we want a society with good values around us.

Finally, we are so proud to live in a society that "reinvents itself" in every generation. I think we should put more effort in it, so that the older generation may and should hand over some of the good values to the new one. I don't have any problem knowing that each generation has their own new set of values that are different than the older ones had, but if one tries to picture a society where all the values will change with the change of every generation, then one needs to understand that the fast changing value system may not work.

I think it is time to depart from the analysis of "Value System" and move on to analyze philosophy as the science of the sciences. I realize that I didn't build a "Complete New Value System" for humanity (I think that is an impossible task for one individual), but I think I accomplished opening some minds for many different possibilities and set a few pointers in motion, so that may be worth thinking about. "Value System" has it's own "Directional Process" just like any system in life, and because of the logical relativity of "right" and "wrong" in time, one could only try to enhance the existing system, with an understanding that the process will go on for generations. I think as long as we design a **"Value System"** based on the "Artificial, Unequal, Fair Share Distribution System", the "Attributes of Human Beings Theory" and based on the "Genetic and Social Inheritance of the Individual", we will be very close to a working, viable, progressive "Value System", that creates a very pleasant and "Free" environment to live in. The better the system design, the closer we get to "Real Freedom".

Part 2 - Chapter 17
Philosophy

Philosophy, Section Five

Section Five, the extended final view of "**Philosophy as the Science of the Sciences**" and the introduction to the concept of the "**Free Institute of Philosophy**". Here I will extend on Chart 1 and I will incorporate the observations of Chart 1 and Chart 26 in the new Chart 49 to create a representation of "**Philosophy as the Science of the Sciences**" in the "**3D + Time Universe**" using Physics as an example.

One needs to understand that the next two charts Chart 49 and Chart 50 are "**concept**" charts to introduce a concept, not charts to describe the "world of Physics at 100%". Following this concept, that would be valid and doable, but here for my investigation it would be too complicated and I do not need that complexity here to introduce my concept.

Let me introduce and analyze Chart 49 and Chart 50, that is reflecting the relation between the sciences (here physics and philosophy), and also explain why I think that "**Philosophy is the Science of the Sciences**".

Part 2 - Chapter 17
Philosophy

Philosophy, Section Five (cont.)

<u>**Chart 49**</u>.   <u>**Philosophy as the Science of the Sciences**</u> in the "**3D + Time Universe**"

t	x	y	z	
t-**R**	x-**R**	y-**R**	z-**R**	-**R** for the "limit" of our -(3D + Time Universe).
.	.	.	.	
t-q	x-q	y-q	z-q	
.	.	.	.	
t-2	x-2	y-2	z-2	
t-1	x-1	y-1	z-1	
t0	x0	y0	z0	
t1	x1	y1	z1	
t2	x2	y2	z2	
.	.	.	.	
tm	xm	ym	zm	**m** for **mechanics**
.	.	.	.	
t1	x**a**	y**a**	z**a**	**a** for average human knowledge of **physics** at **t1**
t1	x**ph**	y**ph**	z**ph**	**ph** for philosopher knowledge of **physics** at **t1**
.	.	.	.	
t1	x**n**	y**n**	z**n**	**n** for best physicist knowledge of **physics** at **t1**
.	.	.	.	
t2	x**a**+1	y**a**+1	z**a**+1	**a+1** for average human knowledge of **physics** at **t2**
t2	x**ph**+1	y**ph**+1	z**ph**+1	**ph+1** for philosopher knowledge of **physics** at **t2**
.	.	.	.	
t2	x**n**+1	y**n**+1	z**n**+1	**n+1** for best physicist knowledge of **physics** at **t2**
.	.	.	.	
t**p**	x**p**	y**p**	z**p**	**p** for **maximum knowledge of physics**
.	.	.	.	
t**c**	x**c**	y**c**	z**c**	**c** for **chemistry**
.	.	.	.	
t**b**	x**b**	y**b**	z**b**	**b** for **biology**
.	.	.	.	
t**ps**	x**ps**	y**ps**	z**ps**	**ps** for **psychology**
.	.	.	.	
t**s**	x**s**	y**s**	z**s**	**s** for **sociology**
.	.	.	.	
t**i**	x**i**	y**i**	z**i**	**i** for **interactions of intelligent species**
.	.	.	.	
t**R**	x**R**	y**R**	z**R**	**R** for the "limit" of our 3D + Time Universe.

Cx (bracket spanning from t1 xa row down to tp xp row)

Part 2 - Chapter 17
Philosophy

Philosophy, Section Five (cont.)

Chart 50.   **(Chart 49 cut out in the middle to investigate)**

tm	xm	ym	zm	**m** for **mechanics**
.	.	.	.	
.	.	.	.	
t1	xa	ya	za	**a** for average human knowledge of **physics** at **t1**
t1	xph	yph	zph	**ph** for philosopher knowledge of **physics** at **t1**
.	.	.	.	
t1	xn	yn	zn	**n** for best physicist knowledge of **physics** at **t1**
.	.	.	.	
t2	xa+1	ya+1	za+1	**a+1** for average human knowledge of **physics** at **t2**
t2	xph+1	yph+1	zph+1	**ph+1** for philosopher knowledge of **physics** at **t2**
.	.	.	.	
t2	xn+1	yn+1	zn+1	**n+1** for best physicist knowledge of **physics** at **t2**
.	.	.	.	
tp	xp	yp	zp	**p** for **maximum knowledge of physics**
.	.	.	.	
.	.	.	.	
tc	xc	yc	zc	**c** for **chemistry**

Cx (bracket spanning the rows)

In this Chart 50, "t" is the time coordinate, "x, y, z" are the three coordinates of the three dimensional universe. "R" represents the "Real Numbers" from mathematics (the one to one relation between the linear line and a number assigned to each of it's points, and extended from −infinity to +infinity (-R, -infinity; R, +infinity). The next are R=**m** for the maximum knowledge of **mechanics,** R=1 for the **t1 time definition,** R=**a** for the average human knowledge of **physics** at **t1** point of time.

Part 2 - Chapter 17
Philosophy

Philosophy, Section Five (cont.)

R=**ph** for the philosopher knowledge of **physics** at **t1** point of time, R=**n** for the best physicist knowledge of **physics** at **t1** point of time, R=**2** for the t**2 time definition,** R=**a**+1 for the average human knowledge of **physics** at **t2** point of time, R=**ph**+1 for the philosopher knowledge of **physics** at **t2** point of time, R=**n**+1 for the best physicist knowledge of **physics** at **t2** point of time, R=**p** for the **maximum knowledge of physics,** R=**c** for the maximum knowledge of **chemistry. Cx** is a complexity arrow that points from the least complex to the more complex.

One can see the following conclusions from this Chart 50. For one to be a philosopher, it requires to have an above average knowledge of Physics. I introduced earlier that philosophers use their Right and Left Hemispheres on the "little better" than average level. The best physicist with the (possible) use of his or her Left Hemisphere (logical side) has a lot larger knowledge than a philosopher.

We can see that at the t**1** time point, the sequence of knowledge level is **a, ph, n.** We also can see that, for example, in the time line t**1,** (for example 2500 years ago) the best physicist's knowledge **n** could be less than our current average (in 2003) **a**+1, at the t**2** time point. If we look at the chart in the t**2** time point, we can see that the sequence of the knowledge level didn't change, **a**+1, **ph**+1. **n**+1. Now, one can extend this concept to all the areas of the knowledge of life to understand that a good philosopher needs to study hard and has to have a very sensitive observation skill, and good logic to connect all things. That can take many years to develop in one's mind. When someone has an above average knowledge base on many areas of life, things start to connect in the most unusual ways in one's mind.

When I say many areas, I mean for example: Physics, Biology, Outdoor Cooking, Fishing, Swimming, Cigar Smoking, Chess, Dancing, etc.

If one thinks that is impossible, that one is mistaken. For example, if one takes 10 people and ranks them with a "physics test" and the average number is 5, then coming up to 6 tells us that one is above average. (That is very far away, to be the best number 10.) It gets a lot easier if one takes 100 or 1000 people for example. Being 501 out of 1000 is better than average.

Part 2 - Chapter 17
Philosophy

Philosophy, Section Five (cont.)

One can see that a philosopher doesn't have to be a genius, however, one needs to have a "drive" and a "desire" to acquire better than average knowledge in all areas of life. (Having a very sensitive observation skill does it most of the time!) The more one succeeds, the better philosopher one can become. One can also see that the knowledge base always increases in the timeline, so the average gets higher, the best get higher, so a philosopher must get higher.

The way Philosophy can become a Science of the Sciences is by the "Connections in the Mind" from different areas of the knowledge base — this leads to some unusual conclusions, that is it challenges the current thinking of the best scientists or politicians or any area of knowledge leaders, then their logical approval or disapproval enhances and challenges the philosopher to accelerate further.

So they inspire and accelerate each other. That is the effect that makes "**Philosophy the Science of the Sciences**"! One also can see that it is the area where the "conflict" between a current and future knowledge base is coming from. The "best of the science" or "leader of a church (theologian)" or "leading politician" strongly believes that their knowledge is "**The Knowledge**", and also believes that is the knowledge that gave them the authority!

So challenging their knowledge is also challenging their authority! (Galileo said, "Sun is the center!" Church authority said, "Do you think we are stupid? God created this world with the Earth in the center!") That is one of the main reasons why philosophers have never been too popular in their time; some get the acknowledgement after they have passed away. I will give an example of how thoughts are connected in my mind — for example, about the possible "**Gun Law**" issue, that I outlined and recommended in the "Value System", section four of this chapter.

The simple rule of "**the two gun rule**" that I am proposing, can solve all the complicated problems that we make such a "big deal" and have such a big argument about. One will see how the politician will react, and they will think I am stupid, uneducated, and clueless, and they can ignore me.

Part 2 - Chapter 17
Philosophy

Philosophy, Section Five (cont.)

That will work in the beginning, until more and more people will become familiar with my simple, logical argument and then they can't ignore it anymore.

The main reason, why no one wants to touch this issue seriously, is because the "gun handling" or "mishandling" is a **behavior** issue! In this emotionally designed "Freedom for All" society, no one can risk their good paying position (except me, because I don't have any!) to say: "If you pick up a Gun, I will take away your **Absolute Freedom** and I will tell you what you can and cannot do!" Yes, If you pick up a Gun, "**You are not Free to Behave anyway you want**", you "**have to**" follow my two simple rules! (Just like I take away your "**Freedom**" when you "**have to**" stop at the "**Red Traffic Light**"! The rules of the many outweighs the rules of the one! One needs to follow the rules, so the traffic system can work for the many!) Back to the Gun stuff! If one wants to stay out of control and Absolute Free, no problem, just don't pick up a Gun!

I have to say again that "behavior" is a "social inheritance", so we need to control and direct the "young ones" to grow up with proper manners and that has nothing to do with "Freedom"! One is more free knowing the rules then living without them. (Knowing and obeying the rules can keep our "young ones" out of jail, so it can keep them free!) Here, logic has to overrule emotions!

Back to the connection example. I like target shooting and I like the movies, as a form of art. I can say that on both areas my knowledge is a little better than average. One day I was watching a movie where John Wayne was teaching his son (in the movie) the rules of gun handling. He says in the movie "The man (of course it is true for women also!) should use the Gun only on two occasions — one to save his own life, second to get meat (hunting, gets meat on the table for food!). That was the time when it "hit me" that he is right; simple, easy-to-follow two rules, and if everyone would live by it in our country (USA), no one would be killed, like it happens sadly every day in our current world.

Part 2 - Chapter 17
Philosophy

Philosophy, Section Five (cont.)

He (in the movie) teaches his son the proper behavior about gun handling. One can see that we don't need to disarm everybody, or outlaw all kinds of guns, we simply have to change the "Mind", and that will change the behavior, and that will change the gun use. If one wants to argue that it isn't that simple, we should try it, to start teaching this stuff in school in a controlled environment, and then check the results. I am not saying that kids should have gun ownership, they shouldn't, however, they should have early education on gun safety, gun cleaning, and proper usage under supervision by adults at a certain age.

One can see in this example that an afternoon spent movie watching can trigger a philosopher's mind to connect the words from the movie with the gun laws and a possible new concept for the "Value System" change. Of course, one needs to understand that the gun business is a huge business in this country (USA) and a lot of money is in it, to go around. Like many times, when a lot of "bucks" are involved, the interest groups of the "money" make the rules, not the common sense logic. The more complicated, the better for some, so everybody can get their little or big power game out of it, and of course, the money that comes with it. Another example, why a good, working , and fair "Value System" can't be built by the rule of the money!

My next subject matter is the need for a "**Free Institute of Philosophy**". One can see that even in our country (USA), the freedom of speech runs into some difficulties. Coming from "Communism", I saw it first hand, that in the communist countries, one can get more years in prison there, for speaking one's mind (speaking against the system) than killing another person. Strange as it seems, I learned it early on (in my own experience) what free speech can do to an individual in a not-free society. Here in our free USA, unfortunately, we are moving to the direction of "Free Politically Correct Speech" from the "Freedom of Speech".

Free as long as it is politically correct. In other words (no surprise for my advanced readers), "**Free as long it won't offend any one's Emotion**"! The problem is that reality many times is logically based, and logical events don't care about human emotions.

311

Part 2 - Chapter 17
Philosophy

Philosophy, Section Five (cont.)

For example, an Earthquake that kills hundreds doesn't care about any human life, it is going by the rule of physics, and no human consciousness is involved in the Earthquake's decision!

If we want to debate issues of reality, sometimes that is emotionally very disturbing, and because we are living in a "very sensitive" society, we get offended real quick! I noticed talking to people that everybody is very quick to form opinions and exercise their freedom of speech, but when I respond with a touch of reality 95 out of 100 people get offended! First I thought it is my English, then I realized, that the problem is that people don't like to face reality or live in reality.

One can see my point that we need a creation and founding of a "**Free Institute of Philosophy**", ("Free" means free from the fear of "Inquisition" and the philosophers of that institute can be free to think on any subject matter, any time, and any way) with the purpose to study, organize, and keep Philosophy up to date, and "**recommend changes**" in all those areas that philosophy analyzes, investigates, and concludes about, as the "sciences", "social structure" of society, and all the areas of our life, **when the design becomes outdated** or **when the directional theory calls for it** (at least once in every generation of life, about 25 years!).

I argued before that I think we need a well designed simplified system (like traffic control system) to live by instead of a non-working "Representative Democracy". (Aristotle concluded about 2500 years ago that democracy doesn't work! How long do we want to keep trying?!) If one wants to challenge me, just answer this, Why is our legislative branch not finding solutions for Health Care Reform?, Social Security Reform?, Tax Reform?, etc? (Because they are above it, so it isn't truly important for them!)

We need a "New Simple Working System" that can be recommended by the "**Free Institute of Philosophy**" and implemented by the government. This free institute can inform the Government (as the Guardian of the new system) of the problems and needed changes.

Part 2 - Chapter 17
Philosophy

Philosophy, Section Five (cont.)

Then the Government can interview it's citizens, (for example, through survey's) collect the information that is needed to propose a possible better "Value System" for all the Individuals for consideration and for the decision making vote!

It can be iterative, multiple votes, and corrections until the final "New Value System" passes with at least 90% of the votes of all! This can create a real checks and balances system, because the original plan, planned by our founding fathers, was gone a long time ago. Today, the two branches are formed from the same wealthy individuals (with a few exceptions), and they are from the same team. One needs to understand that all these changes are needed for the assurance of the survival of the human race, with the leading force of the USA. (I hope!)

Part 2 - Summary
Summary of Thoughts

Summary of Thoughts

Thank you for reading my book this far. I hope all of you enjoyed this "different" type of journey in life. Different in a sense that it took place in our minds. I always knew that we (humanity) are capable of doing amazing things, when our minds and resources are focused on "one major goal". **The survival of our human race should be that goal!**

I am sure that in this book many of you found some good, bad, disturbing, impossible, futuristic, etc., ideas. One needs to understand that those are words to describe some of our human emotional feelings. Nothing wrong with emotions, as I said that many times before, however, our future has to build on a logical, self-understanding humanity for our survival reason. Many times people asked me if I am a "positive" or "negative" person. My answer was (for their surprise) that I am a "realistic" person "with a small positive touch".

For example: many of you know the always used (logically misleading) example of a half cup. People say if your cup is half full, you are positive; if it is half empty, you are negative in your thinking. Just like always, the reflection of the human emotions with no reality. In reality, one sees a cup filled with liquid half-way, that is my "realistic" thinking. (One can't know if it is in the process of being emptied out or getting filled up. Nobody can tell that in reality unless one observes the next step!) Then comes my "small positive touch" to think that the next step will be that the "cup gets filled up and it has my favorite cognac in it and someone will run into the room with my favorite cigar in one's hand for me"! (Basically, this example reflects a mixed use of formal and dialectic logic, that many people fall for, based on the lack of clear understanding the logic — in one time moment "formal" or the logic in the continuous passing time "dialectic".)

The most important fact is that I accomplished something if I made all of my readers think with me in this journey. I don't expect everybody to agree with all my points and views, but I am hoping that "intruding" on all of your thinking minds and emotional world, that I may have just started a thinking process in your mind, and that will lead to a much greater result than I ever could imagine writing this book.

Part 2 - Summary
Summary of Thoughts

Summary of Thoughts (cont.)

I am sure some of the arguments of this book, like the "**Directional Theory**", "**3D + Time Universe Chart**", "**Knowledge Theory Chart**", "**Attributes of Human Beings Theory**", "**Fair Share Distribution Systems Theory**", "**Solutions and possibilities for the Future of Humanity**", "**Realistic Logicism**", Value Systems of Societies", "**Philosophy the Science of the Sciences**", "**McKaneism as a philosophy**", and the "**Free Institute of Philosophy**" concepts all introduced some new ways of looking at our existence, by providing some observations of our weaknesses and the possible proper conclusions and solutions for our human future.

Finally, I would like to present a few very personal thoughts about writing a philosophy book. Philosophy is a reflection of our "Human" existence with all its strengths and weaknesses, and it is extremely difficult to write about those.

Writing philosophy is a very difficult task. Philosophy, in this context as my arguments, as the science of sciences, has to search for and organize the truth. We all know by now from this book, (I hope) that the "truth" is very relative. The "**truth is the mirror reflection of reality in our own human mind**" and as we saw before in the "Directional Theory World" the reality is constantly changing, and so is the reflection of it.

Flying wasn't a reality with a 747 two hundred years ago, it is today. Then the many "screen doors" we, humanity as individuals are looking through in 2003, add to the confusion. (Thought processes of religions, etc.) On the top of that, philosophy has a very emotional side to the philosopher, that has to be controlled in the writing of reality. One needs to be able to doubt things, and ask very emotionally uncomfortable questions in the search for the relative truth. Most philosophers are way ahead of their time in the complexity of thinking and observation skills of their fellow human beings, and the logical concluding way of thinking. That creates a conflict, that one can't ignore, because observing it from history, many philosophers were killed or ridiculed for their thoughts being unusual and against their time of current existing thought systems and the power structure of their establishments.

Part 2 - Summary
Summary of Thoughts

Summary of Thoughts (cont.)

In the process of discovering the "truth" the philosopher has to constantly question him or her self. Am I correct with my observations? Am I correct with my statements? Am I correct with the wordings in my language that I use? Am I correct with my logical conclusions? Is there anyone that can get all those things right all the time? The answer is NO! From the relative reality the mission is coming 99.99% close to reality, and if one can accomplish 85%, that is a very high mark! One needs to remember that the discoveries and changes of reality will move on, so does the knowledge of philosophy that tries to reflect it. (Remember Chart 49, Chart 50.)

I experienced first hand, in my observations, that no one likes reality too much. Most people try to exist in their "artificial dreamland", here in the USA (the people in many other countries are not that lucky!, because of their situation, they can't have theirs!), that mostly is created in our own house.

Every home I ever visited was reflecting the dream world of the owner. Nothing wrong with that, but can you imagine knowing this fact, then introducing a "home wrecking" reality thought process for those people? When the philosopher is embarking on an adventure with all the excitement and hope for a fantastic journey, that philosopher wishes to find the pot of gold at the end of the journey (or at the end of the rainbow). One could find that the pot (if it exists?!) contains the benefits for Humanity, and very little or nothing for the individual philosopher.

One can see (after they are finished reading this book) that an individual philosopher always will find oneself existing in the confined boundaries within a society that exists within the confined boundaries and limitation of the existence of the species (In this case Humanity!), instead of finding that "Enlightening Freedom" that the Philosopher hoped to find. The discoveries in philosophy could and can be a very powerful and destructive emotional force in the philosopher's emotional life.

All these bad things, one can say and ask: Then why write philosophy? I don't think I have a solid, well founded answer for that. It had many affecting components to it.

Part 2 - Summary
Summary of Thoughts

Summary of Thoughts (cont.)

It is an "inside drive" in my mind, that I have to tell these thoughts to everybody. Somehow I know?! that humanity will benefit from these thoughts and thought processes and from this philosophy that I called McKaneism.

Interestingly, when I was visiting my son's high school (around his age of seventeen) and I saw the possible trades and professions that someone can prepare for in life, I was amazed by the fantastic varieties of choices that one can make and how much help the office of consulting of the school can provide.

On the other hand, I was also amazed that reading the alphabetical ordered list, "Philosophy" was missing from it. It is still handled like that "kind of stuff" that one can pick up a semester in some university or college for an extra 2-4 credit points. The philosophical publications (even quarterly articles) that one can find in the libraries are unimportant investigations, some of the very small areas of some unimportant philosophy from the past. These facts definitely contributed to my decision to write.

When I came to this country (USA) in 1980, I chose this country to be my "Home". (I am not here because my Mom and Dad couldn't afford to buy a movie ticket and did "something else" instead, I am here by my own choice — thinking it over, deciding it, and making an 8000 miles journey with all the hardship to get here!)

I decided then, that this is the <u>best country</u> (USA) to live in, and if I ever found that things were not going OK here for me or for the society, I would have only two choices.

<u>One</u>, to take a rocket and leave this planet Earth for good (one can't find a better than the best on Earth!), or <u>Two</u>, change the "world" around me, that I am in. When I talk to different individuals and they are telling me sentences like "I never thought about it like that" or "If you put it that way that makes me think differently", then I made my decision to choose the second choice.

At one of the most joyful events in our human existence, is the question comes up at a very important time of a person's life: "Please speak now or forever hold your peace!".

I did speak my mind here and now as of the year 2003, and now I can "forever hold my peace"!

317

Part 2 - Recommended Readings

List of Recommended Readings

List of Recommended Readings

Aristotle

Selections

Aristotle's systematic treatises

Asimov, Isaac

I, Robot

Prelude to Foundation

Foundation

Second Foundation

Foundation and Empire

Baumol, William J.,
Blinder, Alan S.

ECONOMICS Principles and Policies, Third Edition

Brown, Tom Jr.,
with Morgan, Brandt

Tom Brown's Field Guide to Wilderness Survival

Coser, Lewis A.
Nock, Steven L.
Steffan, Patricia A.
Spain, Daphne

INTRODUCTION TO SOCIOLOGY, Third Edition

Deshimaru, Taisen

THE ZENWAY TO THE MARTIAL ARTS

Dr. Rademacher, Robert A.,
Dr. Gibson, Harry L.

An Introduction to Computers and Information Systems

Part 2 - Recommended Readings
List of Recommended Readings

List of Recommended Readings (cont.)

Einstein, Albert	*Specific Relativity*
	General Relativity
Engels, Frederick	*Poverty of Philosophy* and *Manifesto*
Euclid's	*Elements*
Freud, Sigmund (by Strachey, James and Freud, Anna)	*Standard Edition of the Complete Psychological Works of Sigmund Freud*
Hegel, G. W. F.	*The Phenomenology of Spirit*
Hirai, Tomio M.D.	*ZEN AND THE MIND*
Hoyle, Fred	*The Black Cloud*
Joyce, James	*Ulysses*
Kafka, Franz	*The Metamorphosis; The Trial*
Kant, Immanuel	*Prolegomena to Any Future Metaphysics*
	Critique of Practical Reason
	Fundamental Principles of the Metaphysic of Morals

Part 2 - Recommended Readings
List of Recommended Readings

List of Recommended Readings (cont.)

Kim, Ashida	*NINJA MIND CONTROL*
Lee, Bruce	*TAO of JEET KUNE DO*
Lem, Stanislaw	*His Master's Voice*
Lenin, Vladimir I.	*State and the Revolution*
Lukacs, Gyorgy	*Estetika* (Hungarian)
Marx, Karl	*Capital (Vol. I, II, III)*
More, (Sir) Thomas	*Utopia*
Nitobe, Inazo	*BUSHIDO, The Warrior's Code*
Pavlov, Ivan P.	*Lectures on Conditioned Reflexes*
Plato	*Great Dialogs of Plato* *The Republic of Plato*
Proust, Marcel	*Remembrance of Things Past* *(A la recherche du temps perdu)*
Rejto, Jeno (P. Howard)	*"All his Works"* (Hungarian novels)

Part 2 - Recommended Readings
List of Recommended Readings

List of Recommended Readings (cont.)

Sartre, Jan Paul   *Existentialism and Human Emotions*

Shakespeare, William   *"All his Works", (Comedies, Dramas, Tragedies, Poems)*

Shanks, Bernard   *Wilderness Survival*

Shelly, Gary B.
Cashman, Thomas J.   *BUSINESS SYSTEMS ANALYSIS AND DESIGN*

Smith, Huston   *The World's Religions*

Stumpf, Samuel Enoch   *Socrates to Sartre, A History of Philosophy*

Tohei, Koichi   *KI IN DAILY LIFE*

Ueshiba, Kisshomaru   *Aikido*

Zweig, Stefan   *The Royal Game*

About the Author

Frank P. McKane, the author of "McKaneism - The Foundation of Humanity", has first hand experience by living in the societies of Europe and the USA. These experiences and observations, along with a deep interest of Philosophy, inspired him for a more than thirty years journey of studying, observing, and finally discovering his own philosophy. The long term interest and commitment to analyze and understand humanity to the fullest possible extent, founded the thought processes of the author that one can find in this exciting publication. The realistic and logical thinking of the author creates this well rounded, complete, and very serious philosophy book, with the intention to guide the thinking of humanity into and through the next millennium.

www.ingramcontent.com/pod-product-compliance
Lightning Source LLC
Chambersburg PA
CBHW081106170526
45165CB00008B/2341